a biography

NOT FOR NAUGHT

humble soldier, remarkable life

ARTHUR LEE BURNS

To my wife, Mary.
To my three kids, Lori, Jackie, & Tommy.
To my family.

And to the many heroes of war,
whether you were an infantryman, a baker, a cook,
or a candlestick maker,
you were and you are my hero.

Table Of Contents

FOREWORD

So, it goes…
The kids said I should write a book.
To which, I remember my reply, *"…Like…an obituary kind of thing?!"*
With lots of accolades, they assured me it was not *that* kind of thing.
I'm only 93, after all.

But let's get something straight. I do not, under any circumstance, consider myself to be a remarkable person.

"I think you're remarkable, Artie," says Mary.

No.
In fact, I consider myself to be the most average guy that ever walked the face of the earth!

However, I did live in remarkable times. Some of my associations or occurrences, sure, they were probably 'remarkable.' Memorable. Unordinary in some cases and weird in others. Different or maybe not the norm. Some were even dangerous. Yet, it's been a good life.

So, here are a few of the people and things that I remember, with a little help from my family. The story of an American boy who lived in some remarkable times, having transitioned between horse and buggy days to space travel. And how very fortunate I am to live in such a time as this – period.

Here it goes…

HUMBLE BEGINNINGS

It was October 1930 when I was born, shortly after the beginning of The Great Depression, a historical period of economic turmoil whose effects spanned the entire world. It was a formative time, one that established the very values of my life to come.

I was told that I was delivered by a local doctor in the comforts of our Boyceville, Wisconsin farmhouse. This doctor happened to be of the Jewish faith and, as such, convinced my parents to have me circumcised. According to the doctor, and his faith, it was best to have this procedure on the eighth day after birth. They say at that time a child hasn't fully developed his or her pain recognition. And I guess everything went well because I didn't feel a thing!

Okay, fine. The very first thing I really *do* remember, my earliest memory if you will, is sitting down on the ground, outside a house. My mother was sitting with me, and my father came out a door to join us. Years later I would share with them this detailed image and they would say, "...*But, you were too young...*" because what I described was our house in Boyceville, Wisconsin. Certainly, it was just one of those strange oddities, remembering something so specific, so young, 2 years old or less.

The Great Depression would last almost the first decade of my life. At the time of my birth, mom & dad were sharecropping on a farm belonging to the Hart family near Boyceville. As the depression continued, the Hart's son lost his job in Minneapolis. Therefore, his return to the farm put my parents out of a job. This was common, unfortunately, as families all over tried to make and earn their way through one of the toughest times in our nation's history. For Mom and Dad, this forced them to move back home with dad's parents, Grandma & Grandpa Burns in Osage, Minnesota, about 8 miles west of Park Rapids, Minnesota.

In the Spring of 1932, we were officially en route from Boyceville to Osage, off to Grandma & Grandpa Burns' farm. On the way, we stopped at a hotel in Little Falls, Minnesota. At the time, I was just a young, blonde, curly haired baby boy. Also, at that time, a young, blonde, curly haired baby boy had just been abducted in New Jersey, an event known as the Lindbergh Incident of 1932. A nationwide search ensued to find 'Baby Lindy,' as they called him, which placed the entire nation on heightened alert for all blonde, curly-haired, male infants. Charles and Anne Lindbergh, Baby Lindy's folks, used to live in

Little Falls, and sure enough, that's where we were on this one day. I fit all the criteria of Baby Lindy's description.

Again, I was only told, that on the evening of our arrival at this Little Falls hotel, my folks, and my older brother, Fernale, were dining inside the hotel's restaurant. Story has it that I was asleep when suddenly, as my family began to eat, all the workers were gone. Vanished. And all the doors, locked.

Out of nowhere, FBI agents swept through the restaurant. Seemingly quite prepared, they approached our table, picked me up swiftly and took me away to a room where they had a scale. Quickly, I was weighed and, of course, I weighed the same as Baby Lindy.

It wasn't until I woke up and revealed my brown eyes that they were able to conclude I was not Lindbergh's missing Baby Lindy. His eyes were blue.

In yet another failed attempt to rescue this poor baby, I was returned to my more than likely startled and anxious parents.

Me, Arthur, as a baby,
maybe two or so, with blonde, curly hair

Grandma & Grandpa Burns
Elmyra & Ed

Grandma & Grandpa Burns were both from the Cape Girardeau area of Illinois, a large community of Scots, Irish, & other Northern European immigrants. After my grandpa completed a job where he worked as a farm hand in Iowa, he returned to Illinois to marry my Grandma. Her name was Elmyra Walston. Ella for short, and to me, Grandma Burns.

Once married, Grandma & Grandpa Burns first acquired a large farm near Zearing, Iowa. After an unexpected family tragedy, they moved to Marietta, Minnesota where they purchased another large farm, the land of heavy clay which required intensive labor. While here, Grandpa drove a horse drawn school bus for the local high school students. Around 1927, they purchased a large farm northwest of Osage, which made them the owners of all the land between the cemetery and Straight Lake.

Grandma Burns was a stern, tall, Irish woman, the second born of twelve children. A very capable, determined, well-educated gal. She always had her own horse and buggy, so she either drove her buggy in the summer if the weather was decent or took a horse drawn sleigh in the winter, like to Park Rapids where she did all her banking.

Grandma was the type that always wanted things to be equal and I think we all inherited that family value. She championed Women's Suffrage, advocating for the women's right to vote. She traveled to various places in Iowa, Southern Minnesota, & South Dakota to meet with other women and give speeches. She was also very active in The Temperance Movement which was dedicated to the abstinence of alcohol consumption. Grandma came from a big Irish family with lots of heavy drinking. She had no tolerance for such things.

Grandma did not show much affection, always to the point. But one time, when I was about 6 or 7 years old, she picked me up and put me on her lap and hugged me. Like, *really* hugged me. She talked to me and rocked me back and forth. She just held me for a while, ya know? But no sooner, she must have thought she was being too soft because she put me back down and said, "Okay, that's enough." In that moment and from then on, I knew she loved me.

In the summers, I stayed with her and Grandpa so I could attend bible school at this church in Osage, about a half a mile away from their farm. Grandma got me up in the morning, I'd walk myself to church, and later walk myself home.

Living with Grandma though, she was business all the time. One day, I was lollygaggin' around in the morning before school. I took my sweet time until I heard her yell out to Grandpa to get her buggy ready. Sure enough, he got it ready and pulled it out front. She plopped me up in the seat, sat down beside me, and whipped that pony down the road. She made sure to hit every mud puddle she possibly could, slammed up to the church, and practically threw me out of the carriage. "THERE!" Grandma said. There wasn't any time in her life for that kind of lollygaggin' nonsense.

Another example of her no-nonsense attitude can be illustrated by the following event. One of my cousins was notorious for throwing outrageous tantrums. He would lay on the floor kicking and screaming 'til he got his way. While visiting one time with Grandma Burns' and the rest of us, he threw one of these tantrums. So, calmly, Grandma went and pumped a pail of fresh, ice-cold drinking water and poured it on him. I never heard of him pulling this trick again. At least not at Grandma Burns' house.

Grandma Burns

Ed Burns, Grandpa Burns to me, was the 4th born of nine children. A great man, strong, hard-working, and very capable. He only completed up to the third grade, so technically he was illiterate. He was a workaholic. And physically, he was hard as a rock.

Grandpa grew up during difficult times. He told me many stories of fights he was in, just to exist. I got to witness his last confrontation. One day, after he'd retired, I walked with him from his home in Osage to the post office, which was located inside a large grocery store called Sartain's Store. The owner & postmaster, Mr. Sartain, was a tall man, taller than Grandpa anyway. When we walked in, "Hi Ed," said Mr. Sartain, and slapped Grandpa's hat right off his head.

Grandpa hit him back with one punch. Right into a stove which then knocked down a pipe chimney that ran the distance of the store. Soot and ashes rained down from above and all over Mr. Sartain. Grandpa stepped through the post office door for postal personnel, picked up his own mail, and took my hand as we left the premises.

On the walk home, Grandpa said to me, "Don't *ever* let a man knock off your hat."

Yet, Grandpa was also soft, artistic, very talented, and incredibly inventive. He made inlay in guns using ivory from piano keys. He engraved cutting knives using brass, lead, and copper. And whatever he created was just beautiful. He was also great with horses which were key in those days.

He was a musician, too, played the fiddle, and my grandma, the organ. They played at church and revival camps in the '30s & '40s. I had the pleasure to go with them a few times when I was younger. Most evenings at their home, they played and sang together in their living room, the room dimly lit by their kerosene lamps.

Here's a great story...

Grandpa had this habit. Any penny he was given in change, he threw away. After a while of this habit, Grandma, being the frugal and wise one, finally said to him, "*Please*...give *me* the pennies." So they had a deal. From then on, Grandpa gave all the pennies to Grandma instead of throwing them away.

One cold winter, in the heart of The Great Depression, they ran out of coal to heat their house. Grandma counted up her pennies and turned out, she saved enough pennies to buy one ton of coal, which was enough to heat their home for the rest of that cold winter. Needless to say, Grandma decided early on that she would oversee anything business and financial.

Grandpa Burns
with his beloved Percheron horses

Farm Life at The Burns' Farmhouse
Osage, Minnesota

On the main level, the farmhouse had a bedroom, large kitchen, dining room, and living room. Upstairs I remember looking down a long hallway with many doorways. In fact, there were six bedrooms.

The bottom of the stairway led right into the big dining area, a space I would call, 'Fellowship Hall,' where we ate all our meals and socialized. There was also a screened summer kitchen with an old wood and coal-burning cook stove.

Barley, oats, and hay were the principal crops grown on the Burns' farm. They had a few cows for milking and always had horses.

In the winter, Grandpa and his sons sawed their own ice out of the river and drew it by horses to the big icehouse they had. When the local vendor ran out of ice one summer, my grandpa, smart enough, sold them his blocks of ice. Even in the winter, no one stopped to relax.

I especially remember Threshing Day on the Burns' farm, usually late August, or September. For most, if not all, farmers & families, Threshing Day was the greatest, biggest, most important day of the year. Exciting, too, even for the kids. It was harvesting time!

About a day ahead of Threshing Day, I could already see down the field, the smoke billowing from the steam engine that powered the threshing machine. This monster of a machine would move slowly but surely in our direction, as loud as a train engine, all the way from the Ponsford Prairie region of Minnesota.

On the actual threshing day, the neighboring group of farmers that was part of a collective unit came with their teams of horses and wagons. The thresher operator contracted with approximately twelve farmers. The grains were collected through a long chute and then the farmers would load their bundles of grain into their wagons, in their own specific way. They drove up to the long, elevated conveyor belt, one on either side of the steam powered thresher, and little by little, they threshed all the grain, which was taken to either a grain elevator or storage granary.

Each farmer had to cut and bundle his grain, six to eight bundles to a shock. Drying took at least two weeks which is why, above all, during that drying period, the farmers hoped that it would…not…rain.

Harvest time was exhausting work! It would usually take one long day to thresh 200 acres. With set up and all the farmers in the area on site, it could take about 20 days to a month to complete the threshing on the Ponsford Prairie.

Threshing Day! A very big day, indeed. Straw hats were worn, and the women cooked all day, making big lunches and dinners for everyone working hard on the farm. *Man,* what a time to remember.

The only photo that exists of the Burns clan.
Top left to right: Dora, Bonnie, Grandma Elmira, Maude
Second row: Edward Levi, Boyd, Grandpa Ed
Third row: Loren, Jim, Dale
Front row: Louise, Laurence, Fernale, me
The structure on the right is the summer kitchen that was attached to the main house in Osage.

Grandma & Grandpa never owned a car, nor did they ever have electricity or running water. It was all old time living for them. Even in their later life, when they moved into town and sold their last farm, they built their own house and still didn't opt for running water or electricity and that's how they lived the rest of their lives. They just didn't understand it, primarily, and did not want to try and learn.

"Why try and get electricity when you have kerosene lights and go to bed when the sun goes down?"

That was just their thing.

Plus, Grandma always cooked with a cookstove, using wood or coal. And at night, to softly light a room, they'd just turn on a kerosene lamp, play some music, and go to bed. So, *why would we need electricity?* And the well and pump was right outside so *who says we need running water?* It was just that kind of thinking.

Grandma was 80 years old when she first experienced electricity and other modern conveniences in her Park Rapids apartment.

All the way up to her last moments with us, Grandma's mind was just as sharp as could be.

I was particularly fond of my Grandma & Grandpa Burns and enjoyed spending as much time as I could with them.

Grandma & Grandpa Burns
Ed Burns & Elmyra 'Ella' Walston

Grandma & Grandpa Olson
Emelia & Joseph

Joseph Olson, my mother's father, emigrated to the US from Östervallskog, Värmland County, Sweden in 1868 when he was two years old. He had a newborn brother named Olav who did not survive his first year. My mother's mother, Emelia Koselia Christianson, her parents emigrated to the US from Kvinesdal Vest-Agder, Norway, & settled between Millbank, South Dakota and Nassau, Minnesota but on the South Dakota side. Now, they *thought* they were in Minnesota. It wasn't until later, when the state lines were established, separating the Dakotas from Minnesota, that the Christiansons discovered they had been in South Dakota all along.

Emelia Koselia Christianson was born five years after her parents arrived in America in Manitowoc County, Wisconsin, 1877. She was the second born of all her siblings.

How Grandma & Grandpa met, I do not know. But in 1899, Grandma Emelia married Grandpa Joseph.

Grandma & Grandpa Olson
Joseph Olson & Emelia Koselia Christianson

I didn't have as close a relationship with Grandma & Grandpa Olson as I did with Grandma & Grandpa Burns. Where the Olsons lived in South Dakota was about 200 miles away from my home, whereas Grandma & Grandpa Burns were only eight miles away.

Nonetheless, I remember our visits to Grandma & Grandpa Olson's farm in my younger years. My dad would take my mom, sister, and me to the train station in Detroit Lakes, Minnesota. Trains were a common and inexpensive way to travel in the 1930s as trainlines connected most towns of any size. So, we'd take the train to Fargo, North Dakota and then south to Nassau where Grandpa would pick us up. We'd visit for about two weeks at a time and then Dad & my brother would drive down to get us & take us home. Their arrival was always delightful after the long separation.

On one of my visits to the Olson farm, some neighbor ladies met with my grandma to make quilts…for two days straight. This was certainly not very exciting for a seven-year-old boy, so I got under their table and thumped my head underneath. And let me tell you, Grandma Olson did not like that, so she hit me with a pin right on the top of my head. I then proceeded to cut a flower out of her tablecloth to get even with her.

After lunch that same day, I was still upset so I decided to run away. All the way out in the Dakota Prairies & all by my 7-year-old self, I walked about a half a mile down the road and crawled into a water culvert, telling myself for the next four to five hours that I wasn't appreciated and that I was unloved. I stayed there *all* afternoon, thinking and dreaming of how worried my mom, grandma, and all of her quilting ladies must have been, feeling so sorry they had been so mean to me. I thought and dreamt of the search parties that were probably out looking for me. I had officially gone, I thought, and I was going to make them sorry.

And then I got hungry.

Fine, I'll go home, I thought. But when I got there, everyone acted as if I had never been gone at all! Not the reaction I was hoping for. I figured I might as well stay there where it was warm and the meals were great.

One of my fondest memories of Grandpa Olson is of when he taught me how to play chess. He was a very good strategist. Many of the strategies he taught me, I used, and many of which brought victory to my side of the table. The sad part of this story is that about 10 years after he taught me how to play, I beat him.

"Shau!" Grandpa said as he knocked all the pieces off the board. Grandpa played to win and did not accept defeat easily. While that same competitive spirit is in me as well, we never played chess together again.

Grandpa Olson also had a severe hearing problem, probably legally deaf. I noticed family had to speak very loud around him. When I was young, my mom explained the problem to me as best as she could, but it was still kind of alarming for me.

Another fond memory of the Olsons was on their threshing days. Different from Grandpa Burns, Grandpa Olson drove cars and trucks from the very beginning. On threshing day, the farmers would load grandpa's grain truck and I was able to ride with him all the way to the grain elevator in Nassau. After unloading, Grandpa would drive to the bar and have a beer while I waited for him outside. Children were not allowed in, which meant that afterwards, Grandpa'd take me to the grocery store candy counter for a treat, so it always ended up being worth my wait. I usually chose lemon drops.

Grandma Olson was an outgoing, alert, humorous woman. Grandpa Olson could be described as a genteel man and reserved, intelligent, but also humorous. Above all, Grandpa Olson was a kindly man, & Grandma Olson made a perfect match for her husband.

Grandpa & Grandma Olson
Retirement in Park Rapids

My Parents
Edward & Dora

Edward Levi Burns was born on the farm in Zearing, Iowa in May 1903. He was the youngest sibling of four, with an older brother, Boyd, and two sisters, Velva Leota & Maude.

Tragically, Velva Leota contracted cholera and died at the age of nine. The family was then quarantined and, as a really horrific consequence, could not attend her funeral. This was the unexpected family tragedy that preceded the family's relocation from Zearing to Marietta.

Dora Josephine Olson was born in Millbank in 1901. She was the second child of six. Four sisters, Mabel, Cora, Janet, & Florence, and one brother, Arthur. My mom would grow to love horses and later had an offspring of the famous harness horse, Dan Patch.

Mom & an offspring of Dan Patch

Dad & his pet lamb

My mom & dad, Edward Levi & Dora Josephine, met at a dance hall near Nassau in the early '20s and were married in Lac Qui Parle County, Minnesota in 1925.

My Parents
Dora Josephine Olson & Edward Levi Burns
Wedding Day, May 1925

While living in Osage with my Grandma & Grandpa Burns, my parents met Dr. Donald Houston, 'Doc Houston,' well known in Park Rapids. He told my parents of the farm he owned just west of town and offered it to my parents…with some conditions. Previous renters had just destroyed the farmhouse. So, under the conditions that they repair the house, pay the taxes, and prepare a garden space for Doc each spring, they'd have a deal. And that they did.

I remember the farmhouse was, in fact, very dilapidated, just as Doc described. In one of the rooms upstairs, the windows were knocked out and the previous tenant had started raising chickens…inside the house! I'm sure I gave Mom & Dad a few looks or even a few of my words but they assured me that it would be okay. One day.

Over the next few months, they fixed it up and we moved in, just as the leaves began to fall.

Dad chose to make our farm a dairy farm which would require a sizable barn to house the cattle for much of the year. He cut down some large trees & hauled them to a local sawmill. Between him, the lumber he cut down, and a few volunteers, a large cattle barn was built and with an attached hay shed. They then converted a pre-existing shed into another barn for our horses and cows for that first winter. Dad also had to build us a new outhouse, which was a standard in those days, because the original was not salvageable.

Our Family Farmhouse
Park Rapids, Minnesota

At our farmhouse, the upper level had three bedrooms while the main level of the house had a living room, one bedroom, and a large combination kitchen & dining room, what they now call 'open concept.' Back then, a boiler sat on each end of the kitchen stove. We were always able to maintain a reservoir of warm water to wash our hands, clean up, and take our baths. I took mine on Saturdays.

By the summer of 1934, things were much improved at the farmhouse. I mean, we had our own farm! We had milk cows and chickens and the crops in the fields looked promising.

On a personal basis, things were really good, too. As the baby of the family, I was the object of attention and lacked nothing that I can remember.

One day that mid-August, Dad said I was going to spend time with Grandma and Grandpa Burns. I thought to myself, *how good could things get?*

I had been with Grandma and Grandpa Burns for a couple of days when everyone started saying things like, "How would you like to have a baby sister?" I don't remember knowing what a baby was except other people brought them over to our house occasionally, all bundled up, and we all ooh-ed and aah-ed, but then the baby would leave.

A few days later, my father came to pick me up at Grandma & Grandpa Burns.' He was beaming when he said how lucky I was to have a sister. I wondered what he meant.

Upon our arrival home, we walked in the house, and sure enough, there she was in mom's arms. I remember Fernale smiling and looking so smart about it all. Mom gave me a quick hug but then she was back to the baby…

Nobody even cared about me! I began to realize that my wonderful world as the youngest of the family was crashing down and would never be the same.

It took a couple of years before some recognition of me returned.

They named her Joanne Elizabeth, which I thought was a pretty long name. The family soon normalized, and I loved my new baby sister, Joanne, very much.

Farm Life at Home in Park Rapids

Our farm was a small farm, about 160 acres. Mom & Dad felt that no matter what, my siblings & I should grow up to be farmers. Farming and animal care, they knew well, so as we were growing up, that's what we would learn. But neither my father nor my mother were 'teachers,' really. They thought we would all just learn by watching them and by osmosis or something. Therefore, we were to watch them and study them, gather and absorb all the information that we could and that was that.

Naturally in those days, if you were a farmer, you were harvesting grain, shucking corn, and putting up hay. For us, hay was stored in a portion of our big barn so we could easily fill troughs that were in front of the cows.

We learned how to raise calves, pigs, chickens, horses & we even learned how to raise goats. Our instruction included things like feeding the farm animals correctly and why. How we had to be careful when milking the cows, making sure that we safeguarded the milk. We had anywhere from 10 to 18 cows that were milking at any one time.

We learned that when you have a dairy farm, you have to milk the cows in precise 12-hour intervals. If you milk at six in the morning, you have to milk them again at six in the evening. Milking was done all by hand which took some time so, the earlier the better. Since the farm was a distance away from the house, we'd walk back and forth, one or two of us holding a gas or kerosene lamp to light the way.

Mom & Dad also taught us how to separate the cow's milk from the cream, which was done on a hand-cranked separator machine. Once separated, the cream was kept in these big cans and stored in a water cooler in our pumphouse. The cream would then be taken to the local creamery in Park Rapids where they made butter from the cream!

Up until the age of about 8 or 9, I helped Mom with the dishes after our meals while Dad & Fernale worked out in the barn. With every dish, I looked forward to the day I would finally graduate from the kitchen and be old enough to go out with "the guys," when I could trade dish duty for all the much more manly barn duties.

When that day finally came, the day I graduated from the wash & dry routine of the kitchen to all the duties of the barn, just like I had always wanted…boy, was I shocked.

I would very soon learn that I graduated from the kitchen doing dishes, sure, but that's where it was *warm!* Now, I was shoveling every kind of animal poop you could imagine, & in every kind of weather you could imagine, because that's the first job I got!

For quite some time, I thought I had made the most terrible mistake. And when you have an older brother like I did, I got the smaller, less fun kind of jobs at first. It was work nonetheless, and *hard* work at that. It wasn't until I graduated to some greater, more qualified duties, that I felt I could celebrate a little. Looking back, really the best part was just the time spent with my dad & Fernale.

From about 11 years old & on, I was awakened at 5am. During the school months of the year, I helped to milk the cows and separate the cream from the raw milk. After morning milking, I ate breakfast, changed from barn clothes to school clothes & was off to school. After school, it was back to the farm for the evening responsibilities.

"The Guys"
Fernale, Dad, & me

A Tribute to My Mom & Dad

At home, I always loved working with Dad. We both had similar interests. He taught me how to hunt and fish. He had been a boxer in his younger days for a time, so of course I wanted to be that, too. One of our big, treasured moments together was when we would save up the battery to the radio when the Joe Louis fights came on. We would get the cows milked a little early on fight dates, sit around the radio, and listen to the fight. Oh gosh, he would describe all the things. He would explain everything about the jab and the upper cut and the one two punch. Later on, when I was in Golden Gloves and trying to box, I learned that was not my forte and that it never would be.

When I turned 12, we were in the middle of WWII. And Dad taught me how to drive. At that time, all the cars were manual. And well, for me anyway, the pedals were a stretch! I had to sit on two pillows whenever I drove anywhere.

There was *one* bad incident when I was first learning to drive. Around the farm, we would put up our own fencing, about a quarter mile at a time. There was a device that we hooked on to our car and then you just nudge the gas a little and that tightens all the wire so you can staple the fence to the post and secure it. Well, Dad & I were fencing one day, and *I* was behind the wheel. So, wouldn't you know it, my foot slipped off the clutch, tearing down about a quarter mile of fence that we just installed…*and* the posts that they adhered to. Dad was not very happy with my driving to say the least. He decided that he would have to retrain me and possibly use some bigger pillows.

At that time, the State of Minnesota issued a driver's license to any boy at the age of 12 who lived on a farm. It cost $.25 for my license and was good forever.

My dad instilled in me, subtly, the good things of life. Things to do and not to do. To be truthful & honest with your fellow man and to be kind & considerate of others. He spoke adamantly about the proper treatment of children and about being kind to women. How we should honor women and treat them with respect. He said to me that *women are not 'the weaker sex'* and he was right, because I assure you, they are not.

Dad wasn't talkative, neither was Mom. They didn't express much emotion that I remember. One oddity was always stressed that men don't cry. I had never seen my dad cry except one time, when his own father died. Once when we were out cutting wood when I was pretty young, Dad was throwing firewood into a pile that I was then loading into a wagon. And well, one of the logs came up and hit me in the leg and it really hurt so I cried. My dad said to me, "Ohhh, no son, you can't cry." *How strange is that,* I thought… it hurt!

There *is* a time to cry, I have since learned.

My father was a great dad. He was my friend and there was never a doubt that he loved me. It was an ideal father-son relationship, of which I'm grateful.

My mom, having grown up in a Scandinavian family, believed that from every waking hour until the moment you go to bed, you should be working or doing something. There was no time to be idle, no time for laziness or boredom, because there was always something to do. During working hours, you worked.

Although, Mom & Dad *did* work to extremes sometimes. After all, they had grown up in the major farming seasons, where there wasn't time to talk about being busy, they just were. People just kept working, and for long hours if they had to, until the job was done. When it was time to play, okay, then they played. But as a kid, we could never get away with the sayings of today, like *I'm bored!* We would get assigned a task very fast if we did.

Mom's form of play was in her garden. She loved to garden. And after harvest, canning and preserving. Mom baked homemade bread all the years that I was home. She was just a tremendous cook. Dad and I hunted squirrels, rabbits, and pheasants, partridge, deer, you name it. Mom was able to cook it all and it *all* tasted wonderful.

Mom in her garden

After Mom finished high school, she went to Edina, Minnesota for trade school, where she received her credentials both as a certified seamstress and hat maker. She then would take old clothes that were in good condition, cut them all down, and turn them into something wonderful. She would take old throw away coats from other people, collect them, and use them for one of her creations.

While we were young, Mom made a lot of the clothing for us kids. In fact, she made my very first suit from my dad's wedding suit. Everything Mom made was immaculate, as if it was store bought, but better.

Me, wearing the suit my mom made by hand

Mom never drove a car, so I'd drive our car & take Mom and her lady friends to their house parties and church parties and infamous weekend-long quilting extravaganzas. I know Mom & her friends were probably pretty nervous at times when I was behind the wheel, but I got them there and back just fine! We drove tractors and everything else at that age, so it wasn't like I was inexperienced…

One of Mom's flaws was that she was never on time. I remember we'd go to church and were always last to get there. Mom, Dad, Fernale, Joanne & I would make our entrance, put our heads down a little, and walk way up toward the front where all the empty seats were.

Mom? She thought nothing of it and it never phased her. It was almost a revengeful thing she was doing, like, *I'll show 'em! I'll come in when I'm good and ready!* Makes me laugh still to this day. I remember one time, my dad said to her, "You've got a doctor's appointment, you're scheduled to be there, therefore we must go!" Mom barked back, "My time is important too!" It was just a little glitch she had in her personality, but stubbornly sweet about it, too.

She was a very righteous woman. Honest, truthful, & never complained about things. And we, kids, could never complain either, especially about food. That was another one of her 'isms.' Otherwise, it was always the same thing, *they're starvin' in China, ya know!* Around this time, not only were we trying to survive through The Great Depression, but we were also just coming out of the Dust Bowl era where food was short in America, so if we complained about our food, we would quickly be reminded that there were people starving in China. Really, it was just a time of great distress, so Mom took advantage of that and struck it home whenever she needed to. Even if we just happened to say, ya know, "I don't like this," or something, – but first off, that was a no-no to say that. Secondly, Dad always backed her up on it. "You should be thankful for your food!" he would add. And then we would get the whole sermon…with the finality that someone was starving in China.

The food was always very good though, so very seldom do I ever remember my siblings complaining about it and not me either. We were corrected quite rapidly, if so.

My folks always had a small bottle of brandy in the house. If we got sick with the flu or, say, a bad cold, we'd get a little bit of brandy in lemon juice and hot water, menthol (Vicks) on the chest, throat, and bottom of our feet, tucked into bed early, and we all just prayed that we'd survive it.

I learned another lesson around this time. There were some sheep farmers just south of us about a mile or two. A bear had started marauding their farm and killing the sheep. So, the local farmers went and got approval from the game warden to track down and kill this bear. The hunters, including my dad, agreed to divide up the bear meat. We were excited and thought we were so blessed!

But to me and my family, we found it very wild in taste & tough in texture. And as it cools, it gets worse. Maybe some people like bear meat but lesson learned, that was *not* a blessing.

It is both true and a tribute to my mom & dad that we were never hungry. We always had food on the table and that, by itself, in the middle of The Great Depression, was a blessing from my childhood that I will never forget.

This is my family.
Left to right,
Me, Joanne, Mom, Dad, Fernale

My mom, baby Joanne, me, Fernale, & our dog, Rex
"We used to tell him [Rex], *you're the best doggie in the whole world,*
& then he'd wag & then he'd shake." - Joanne

Poplar Grove Country School

My first eight school years took place in a one room schoolhouse out in the country, *Poplar Grove Country School,* known to every kid in a two-mile radius as 'Country School.' Everyone walked to country school then. We lived about a half mile away. Some kids had up to 2 miles in the morning and 2 miles back in the afternoon.

The school served 1st through 8th grade, all led by one teacher. Every student, grades 1-8, sat in one room all day, during all eight grades of instruction. That's right, eight grades, one room, one teacher. Instruction started with the lower grades and worked its way up to the eighth graders through the day. The day started with reading then writing, arithmetic next, and English was the last subject.

The upside for when we were in the younger grades was that we kind of got the coming attractions listening to the upper grades once it was their turn. The downside of country school was that the last subjects of the day got the least amount of instruction because time started running out. By the time instruction got to the 7th & 8th graders, sometimes class was dismissed before all the English material was covered.

Poplar Grove Country School,
Teacher, Miss Schmider, with three eighth grade girls
Miss Schmider would end up being the aunt of Joanne's future husband!

I don't remember a book in our house except a Bible as Mom and Dad didn't have much time to read. I remember the fourth graders had this *fabulous* book. It was a geography book. I loved geography & I loved reading. At the end of my third year, I didn't really swipe it…I just took it home without the proper authorization. All summer, I read that geography book from cover to cover. When I got to fourth grade, I had straight A's in geography. I even read about Madagascar. I could tell you *everything* about Madagascar, that was in the book anyway.

We didn't have running water at the school, only a pump and a pail. The huge woodshed outside was flanked by outhouses, one for the boys on one end, and another for the girls on the other end.

In the winter, the teacher would hire the most capable 8th grade boy to come in a half hour early to start the fire and pile in wood for the day. That was part of the winter ritual.

During Christmas, we sang Christmas carols and always held an annual Christmas play, where every student played a part.

In the better weather, we played softball. We made fake guns out of sticks and rubber bands. And we played, 'Tin Can Alley,' where someone hits the tin can, runs & hides, and then the opponent must find & name the one who hid in order to win. We also played, 'Anti-Over-The-Rooftop.' Someone would throw a ball over the school rooftop, 15 feet tall. Ten people total, five on each side of the school. If someone caught the ball on the other side, 100 points. 1 bounce, 75 points, and 2 bounces got you 50 points.

One of the things we sometimes did in grade school was switch sandwiches with someone over lunch, just to get some variety. For me, I usually had two slices of bread with a little butter and jelly. One time, I switched sandwiches with one of the Cloud boys, and gosh it just tasted funny. "What is this?" I asked the Cloud boy. "It's lard with salt and pepper," he replied. Now, I don't know if you've ever eaten a lard sandwich with a sprinkle of salt & pepper but it's not exactly a delight. That would be the last country school sandwich swap of my country school days. *But,* the bad deal was vindicated the next summer when the same Cloud boy taught me how to swim.

Me, 10 years old

'Poplar Grove' Country School, 1943, Grades 1-8
My sister and I are pictured second row, last two on the right.

One of the subjects taught in 7th or 8th grade was Minnesota History. I learned an interesting legend about the native peoples of Minnesota. There was a chief that promised his daughter, a princess, to another chief. The young princess was not in favor of this arrangement and pleaded with her father to cancel his promise. Her Father told her he couldn't cancel his promise to the other chief, but he would postpone the marriage until the last leaf fell from the trees in the area.

The legend continues with the princess asking the trees to keep their leaves. The elm, the ash, the aspen, they all replied that they could not keep their leaves until the new ones arrived because it went against nature. One tree, the red oak, replied that he could do this. So, the vow of her Father was set aside.

To this day, if you check the forests of Minnesota in the winter and spring, you will see that the red oak holds its leaves of the past year until the new leaves appear – wonderful story, eh?

Changing Times
Through The Lens of a Young American Boy

One awakening for me during this time was we were not the poorest of the poor. Most were in the same boat as one another, which was the state of not having much. The depression had really sunk in by the time I was in grade school, but everyone helped each other through it. There was a time that things had gotten so bad for this one particular family that the parents had to farm out their children because they couldn't feed them. Mom & Dad took in one of their boys for a season or two which was a reminder for me that no matter how bad it seemed, we were not the poorest.

The boy who lived with us later found work with the WPA (see below) and was one of the first to be drafted into WWII. Unfortunately, he was wounded in the South Pacific and lost one of his legs. But after his service to our country, he received government education and went on to establish his own business & was successful for the rest of his life.

The WPA, Works Progress Administration, was an employment and infrastructure program created by President Roosevelt in 1935 to help with employment projects for the general public, like construction and roads and things like that. Kind of a forerunner of the military. They even built some of our parks here in Park Rapids.

Prior military men organized the work groups and set up the rules for the different camps, like, no fighting, no talking when you ate, and no smoking when you were in certain areas. They taught about basic hygiene, the concept of brushing your teeth in the morning and at night, and there was medical aid provided.

As the country entered WWII, this program was a benefit for the mobilization of our military forces.

I became more and more cognizant of the times we were living in. A few of my older cousins also went to work for the WPA in 1939.

The WPA afforded work for two of my other cousins who pooled their money made, and after two years, were able to buy a car and head off to the west coast. Both were very successful in the lumbering business in Oregon until WWII started and they were drafted into service.

I took notice of these things & thought, *things must be getting better, right?* Older children, like these cousins of mine & others from my area were able to go out and find *work*, which was no small feat. And although any income received would be shared to some degree amongst the family, whatever was possible, nonetheless this was a tell-tale sign of progress. That things could quite possibly be looking up.

At home, not only did we not have any books, but there also wasn't a newspaper either. Right after Pearl Harbor, December 7, 1941, I pleaded just enough for a newspaper that Mom & Dad started getting a daily paper in the mail.

This was also when the radio was just coming into its fullest. Now not only did we have the newspaper, but we had the news too, *every* night. That was a big thing. What was happening on the war front was the first segment on the news & was always oriented that way. And now, every movie that came out played at least one clip of the war, too.

Once we started getting the newspaper, I took it, and then using a map, I would map what I read & learned about the wars. I'd do this in my spare time. I'd keep track of where all the naval units and the armies were throughout the world. I kept up with their every move, every battle, and every victory, albeit 2 or 3 weeks behind the happening.

I had always loved geography, places, and situations in the world. I wanted to see, *where'd this army go from here?* News would mention where they bombed someone, *well where? What island?* These were all new names to me, like the Carolinas, *what Carolinas?* I wanted to know.

I imagine I used a pencil, a soft lead pencil, to mark where the soldiers were. I think I kept a journal for a while, too.

Everything I did, all my mapping, was done in our dining room, 'fellowship hall.'

All the guys around me had to move up about 2 years. So even if someone was 14, they were taking jobs like they were 16 because all the 16-year-olds would be gone at 17. Everyone, all of us kids, got a little older real fast. People started getting letters of lost ones in the war as casualties were reported. Between this & the combination of the newspapers and those movie clips, I was impacted.

34

When I turned 13, Fernale was soon going off to the war. Little by little, he & Dad taught me how to speed up milking the cows and how to care for all the animals, giving them hay, food, and water. At the start, I could milk, like, one cow, while Dad & Fernale did three or four. But when Fernale left for war on May 17, 1943, at the age of 17, I was keeping up with Dad on milking and other responsibilities on the farm just fine.

In 1942/43, two new changes occurred at the Burns' farm. My dad got in line as early as he could in 1942 to apply for our first tractor. As soon as it arrived, out went the horses.

In the spring of 1943, the REA, Rural Electrical Association, came to rural Park Rapids and now we had electricity on the farm. Up 'til this time, every Friday night my sister and I had the chore of washing the lamp shades, trimming the wicks, and filling the lamps with kerosene. Now, the world was blossoming right before my eyes! My sister, Joanne would run to the top of the stairs and flip on the light switch, and I would turn it off from the bottom of the stairs. The flicker of light from a light switch? Boy, life was good. We got a tractor, we had electricity, and there was joy in my heart.

For Christmas that same year, Mom & Dad got me a dictionary. For the next year, maybe even two, I read the whole darned thing, every page of that dictionary. I still have it and use it to this day. After that, they started buying me books from the Dave Dawson series by Sidney Bowen.

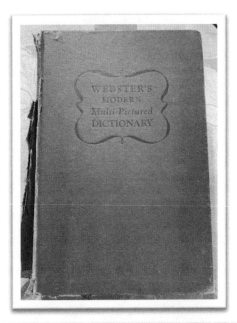

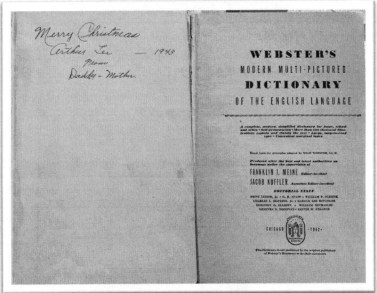

My very first book, Christmas 1943
Webster's Modern Multi-Pictured Dictionary of the English Language, 1942
with my mom & dad's handwritten inscription on the first page

Throughout what felt like no time at all, it became unreal the things we could do. We were able to run the radio all the time. We had power and light in the barn and a yard light, and we had a reading lamp at the dining room table in fellowship hall. We got a pump for water and didn't have to pump by hand anymore.

Collectively, we were all taking note of the little grains of hope that said we were on our way out of The Great Depression. Things were good at thirteen. Things were good indeed, except for the war.

I was *really* starting to take notice of what it meant to be a soldier. Fernale had enlisted in the Navy before finishing high school & was now serving in WWII. Some of our cousins and uncles also enlisted right away. In the window of every house that had someone in the service, there was a flag. They called it the star flag. The stars represented the number of family members in the service. My neighbors, the Jorgensons, had 7 stars on their flag, 5 of them for their sons, and 2 for their daughters. They all made it home with the exception of the oldest son, Arthur, who was killed in the Philippines. Stars on the flag were blue or gold. A gold star represented a son or daughter that was killed in action. So, the Jorgensons had 6 blue stars and one gold.

Then, at one point I read the story of the Sullivan brothers, who all went down on the same ship.

These were things that impacted a young person. Stories of war had officially become a part of my youth and development and deeply affected my psyche.

Star Flag

Park Rapids High School

A year before Fernale left for the Navy, Park Rapids High School burned down, sending all the students to various churches and community centers for classes. This meant that *I* would get to go to the brand-new constructed Park Rapids High School.

I graduated from grade school in the spring of 1944 and began high school in the fall of the same year. What a change. No more walking to school! A bus stopped at our farm's driveway, so I got to ride to school with the other kids that lived within a ten-mile radius. I suppose there were thirty students on our bus each day.

Life was good even at school! I got to enjoy the most modern of school facilities of that time. Bright new hallways and a squeaky-clean gymnasium floor. School desks without chewed gum underneath, not yet anyway. And what I remember most of all was chemistry class, equipped with fresh black laboratory counters and sinks. Not to mention, instructed by one of the best teachers I ever had in my life – Mr. Harold Johnson. He was a brilliant chemistry teacher, a true mastermind of his vocation, who was not only intelligent but took a sincere interest in every one of his students.

Park Rapids High School would also provide me with a treasure; a *library* at the very center of the second floor, grand & new, and two times larger than the entire Poplar Grove Country School. I also had The Carnegie Library just a short walk away. It was a privilege not taken for granted, to be able to check out a book at any time.

The world just opened up for me.

My favorite subject, geography, was no longer taught in high school but replaced with both civics and history, which kept me happy. With my own personal experience of tracking history that was still in the making, I followed along well.

English was my battleground. Since English was always the last subject of the day at Country School, I was really behind on my English and grammar skills. I struggled to know the difference between a subject & a predicate in the beginning. I had to ask for help from tutors and other students in order to make it through high school. It helped that my English teacher took pity on me and helped me on the side with my grammar.

I loved reading immensely and I liked writing, too, so much so that I wrote papers for one of the high school jocks. Until he got a C and complained…that was the end of that gig.

The school had received a brand new 16mm projector for class films, but faculty needed a little help with it when it came time to show different films to different classrooms. I was chosen for the job. Whenever a class was blessed with movie day, I got to sit in and operate the projector, and catch all the films myself. *That* job was not bad.

Outside groups started asking for film projection help after school. The local gun club, the farm club, you name it, I'd show the films at their evening meetings. Quickly, I became known as the tech-savvy guy in school.

But being the "techy" also meant that I would have the misfortune of running the girls' health films…two years in a row. Never saw so many girls blushing at one time. Regardless, it really was a great job.

I also worked at two different gas stations in town, one being Engle's Motors, a Kaiser-Frazer motor company, full service at the time. Two people would usually attack each car. One guy would clean the windshield and check the oil levels, and then if the driver asked, he might also check the air in the tires or the car battery. And of course, the other attendant would pump the gas and collect payment.

On the weekends, I worked setting pins at the bowling alley, which was a manual job back then, and quite the responsibility. You did this with ten holes in the wall behind you and advancing a matchstick in the holes as the game progressed. During league nights, I would work two lanes at once, keeping tabs of which frame each lane was on. Hard work, but the pay was good for an evening's work.

By this time, my schedule each day started at 5am with milking the cows by 6am, followed by breakfast and catching the bus to school. After school I rode the bus home, milked the cows again, had supper, then walked or sometimes hitched a ride back to town to the theater or the bowling alley. League nights were Tuesdays and Thursdays setting pins 'til 11 or 12 at night. Then of course, it was a three mile walk back home in the dark.

One summer, between my sophomore and junior years, I worked as a cabin boy at the Timberlane Lodge on Long Lake, now the Timberlane Resort. Lots of sweaty dock work & cleaning fresh caught fish. I met a lot of wealthy people, including the President of Chevrolet Motor Company who only came to the resort by way of his private float plane.

During my sophomore through senior years of high school, I worked at the theaters in our town. I ushered, took tickets, posted the marque each time the movies changed, and whatever else was needed. I made it to Assistant Manager of one of the theaters by the time I was a senior. There was no TV yet in those years so everyone went to the movies at least once a week. Busy place!

I guess you could say I always had a town job in addition to the farm work back at home. While I was in high school, my father couldn't pay me much. I remember the most I ever got was 50 cents a week. Now, 50 cents in those days *could* buy you a ticket to a show and even an ice cream cone after. But I wanted a little more spending money, ya know? A little more in my pocket. That way, when I took a sweetie out for a date, the least I could do was buy her a soda. Or, maybe I could buy myself a nice shirt somewhere to augment my limited wardrobe.

Mom & Dad were glad to see this. Having been through some of their own rough times during the depression, they thought highly of seeing their son make a living for himself. Plus, I still milked the cows morning and night!

The Gold Rush

In the summer of 1948, the summer before my senior year, I heard tales of making a lot of money working in the canning factories in Southern Minnesota and Northern Iowa and picking apples in the Yakima Valley of Washington. One of my friends, Dean Noland, and I talked this over and thought that we should capitalize on this opportunity and make a lot of money to tide us over through our last year of high school. In mid-June, we set off hitchhiking our way from Park Rapids to Yakima, a 1,500-mile journey roughly.

We somehow got a ride to Grand Forks, North Dakota which was about 140 miles. We stayed in the YMCA for a couple of nights for some reason. I think it was to visit two of his brothers that lived in the area.

We got lucky when we resumed our hitchhiking. It was only late afternoon when a guy headed to Dickinson, North Dakota stopped for us in his new car. He had been on a two-day business trip, so he wanted *us* to drive through the night while *he* slept in the back. We woke him once for a gas fill and arrived in Dickinson the next morning.

Eventually, we checked in at an employment office, but nothing was of interest for a good job.

We were advised to hitchhike south to US Route 10 and follow that out west. Low and behold, a guy picked us up near Belt, Montana, said he owned a ranch about 20 miles from the city, and offered us good pay if we worked there. Away we went.

He had about 1,000 acres of land and raised beef cattle. All work was done by horses, not a tractor in sight. *Bad omen.* We stayed on for about 3 weeks and got the harvest done. We even built and repaired a chimney in the area's schoolhouse for some extra pay. But, after those three weeks, I convinced Dean that we did not go out west to work on a ranch. So, we ended our ranch venture.

The rancher took us back to Belt and we hitched a ride that same day to Great Falls, Montana. No gold mines were in operation, but they had a good job offer at the one and only lumber yard in the area. Pay was good but the costs for housing and food didn't leave us with much. The employment office had said that the apple harvest in Yakima was going to be late that year, so we decided to take the lumber yard offer.

The first duties in the lumber yard consisted of unloading two railroad box cars full of 90-pound rolled roofing and the days after didn't get any better. We must have been doing okay though because they raised our pay after only two weeks. But after paying our living expenses, we still weren't any better off than we were working at home.

By mid-August, we had to get back for school. The lumber yard offered us another raise to stay on, but I knew I had to at least finish high school. Dean & I started our trek home.

We spent a couple of nights in a roadside ditch, not pleasant, in the middle of mosquito and bug season. One of our evenings, we arrived late in a town in North Dakota. We asked, or I should say begged, the Chief of Police if we could sleep the night in his jail. He agreed…but he would have to lock us up, just like everyone else. It was a terrible sound when the jail door slammed shut and he walked away. But we did get a good night's sleep. He must have had pity on us because he even gave us a jailhouse breakfast the next day and drove us outside of town so we could continue our journey home.

At this point, Dean & I had to split up as some people wouldn't pick up two guys in a no-mans-land. It took us both over a week to get home, separately. When I got home, Dean was there eating a good, home cooked meal with *my* parents. He beat me.

I learned many things that summer but here's the moral of the story. First, to really check out the stories of quick gained riches. And second, call the labor office where you're headed *before* you go!

Go Bigger or Get Out

In spite of it all, that I know of, I would become the first male to complete high school on both my mother & father's sides.

When high school graduation was approaching, the economy was not very good. I wanted to go to college, but I didn't have the funds to do so. So, I began to search for what I could do next.

The lifestyle of farming, or the career option of it I should say, was headed towards the idea of bigger acreage, because they were getting bigger tractors and bigger machinery that could handle more acreage. So, you either went bigger or you got out. Nowadays there's all these hobby farms scattered around, which can always be a fun place to have a few chickens or some cattle. But the older I got & the more I knew what was to come of the farming age, I could never see anything but work with a very poor return money wise. And it just wasn't my thing. I quickly determined that I was not going to get in the race. That I was *not* going to be a farmer.

Up to this point, and like many others, we were raised as a nation commending the military. The American soldier was the highest honor. The American soldier was our role model. It was now just a short time after WWII and we were victorious, so we were a very confident population. We believed in God and our country and had a good set of healthy social beliefs. There was a lot of pride in our nation and to serve our nation. Wanting to become a soldier was the norm, it was just the thing you did. Thinking of becoming a soldier myself was a no brainer. Not only was it a viable option, but it was also the best one for me at that time.

In March 1949, my friend, Virgil Zeller, & I went to the National Guard Armory in our hometown of Park Rapids & enlisted. We planned on going into the regular army that summer, but at least this would give us a head-start in military training.

For me, I thought, maybe I could at least go into the military for 3 years or so. Get my training, and then there'd be some schooling available to me afterwards, and I'd see how things went from there. Maybe that was the wisest thing to do. All I knew for sure was that I would be honored to be a soldier & I'd let time take care of the rest.

ARTHUR BURNS
"Artie"
H.S.A.A. 1, 2, 3; "Panther Prints" Staff 1, 2, 3; F.F.A. 2, 3, 4, Treas. 4; Class Play 3; Cheerleader 2; Library Club 1, 2, 3, 4, Vice-Pres. 3; Pep Club 3; Student Council 3; Class Officer 3; Patrol Supervisor 4.

From Farm to Barracks

Guard training consisted of two-hour Monday night drills. And then come the latter part of June, the National Guard had a two-week camp every year, where you go out and train in the field. So that June, we left by convoy in military trucks, and were off to Camp Ripley in Minnesota for our field training.

I guess you could call it basic military training, having included the basics of military discipline. How to salute, how to dress, how to polish shoes. We got paid a day's wages according to our respective ranks. And well, in those days we were called *recruits*. We learned basic military techniques and procedures, how to handle our weapon, which was an M1 Garand rifle from the WWII days. We even learned to put it together blindfolded. We went on the firing range for 3 or 4 days with our assigned weapon and had to qualify with that same weapon.

In my first unit in the National Guard, my platoon leader was one of my cousins that had been in the US Army Air Force as a *bombardier* in the Pacific Ocean theater of WWII. And then another cousin of mine was our First Sergeant who had been in the US Army in the Aleutian Islands during the same war period.

Later that year, on December 5th, 1949, my dad drove Virgil and I to Fargo. We were going into the Army, active duty! My mom gave me a hug on my way out of the house and said *goodbye, behave yourself, be a man, and don't be foolish,* things like that. It seemed like men had gone to war most of my young life, commonplace almost, so me leaving wasn't really anything very dramatic.

It was about an hour and a half ride, 90 miles or so. We pulled up to the big federal building there in Fargo where all the military recruiters, FBI, and other big agencies were located. I took my little suitcase that I had and said goodbye to my dad.

December 6th, 1949, Virgil & I made our way down to the recruiter & were prompted to wait for the rest of the guys to get there. Just a little while after, we and about 20 other men from our area were officially sworn in & were headed to basic training.

They escorted us to the train, which they called the troop train, and we were off. From Fargo, North Dakota to Fort Riley, Kansas. Things were *good*.

Some of the more foolish ones drank a lot of beer on the way down. Virgil & I did not. I don't know if Dad had warned us effectively or if Mom's memo about not being foolish really did work, but when we woke, it was official. We were in *the Army*.

Our train came to a stop in a railroad siding in no-man's land, that I soon learned was part of the US Army's Fort Ripley Complex. It was a cold December day in Kansas, if you can imagine. The wind was coming out of Canada at about 40mph, really just a cold day, when some guy in uniform yelled,

"Get your &$%#^ *&^ outside!"
"You're in the Army now!"
"You better start acting like it!"

They tried their best to march us from the train into this great big building. We had to get rid of all our clothes, put them in a bag and send 'em home. They outfitted us with our basic issue of clothes, which at the time, was just underwear, fatigues, boots, and a field jacket. They gave us our bedding, which consisted of two blankets, two sheets and a single pillowcase. We then carried all of our 'new' items and marched for about a mile to our barracks.

All the way there the drill sergeants yelled things like,

"This is the saddest outfit!..."
"...The saddest group of men we have ever seen in our lives!"

Like, really putting it to us. We got down to our barracks, hurried and made our beds and wouldn't you know it, right then, they let us go to bed...
...For about 15 minutes.

"Get %^& out of bed!"
"What do you think, you're *home*?! And just gonna *sleep* all day?!"

With lots & lots of expletives, we were rushed into the mess hall to eat. Already, the vocabulary was becoming quite challenging. I remember distinctly, one of the cooks,

"How do you like your eggs?"

And I thought I knew better by saying, "Oh, any way you got 'em."
WRONG.

My eggs were all half done and thrown onto my plate.

I also soon learned that toast was not toast. It was either burnt or not toasted at all.

We ate quickly and out we went again. Once more, they attempted to march us the best they could, getting us organized by our respective platoons and squads.

Back at our barracks, they demonstrated how to make the bed the proper way. Then we marched right back to that great big building again to get issued the rest of our clothes and things, which included the rest of our uniforms, shoes, towels, and basic toiletry items. And then marched back again to our barracks. Just another cold, cold day. And demeaning. I couldn't help but think, *what did I get myself into? Do I really want to do this? And for $62 a month? Break that down by the day... — "uff da!"*

On top of all that, we had a really difficult platoon sergeant that slept in our barracks but in his own room. The rest of us were divided into four squads; two upstairs, two downstairs, with rows of bunks. There was one bathroom on each floor so you can imagine, let's see, about 25 men trying to get into the bathroom in the morning and at night. The plus side was that the facilities worked most the time, so we had functioning sinks, showers, and toilets.

Those first few days, though, we arranged our clothes into their proper sequence on their appropriate racks. We set up our footlocker, where our toiletries went on the top & in a *very* certain way. Our towels and our underwear were on the bottom, arranged exactly in certain rows and in certain places. We had a separate place where we could lock up some of our things like gas masks, field packs, and all that.

The next thing they issued was our rifle. Fortunately, it was an M1 Garand, like I had in the National Guard. I remember the serial number still today, 596499, manufactured by *International Harvester,* probably in the early 1940s.

And so, these were the necessities of an infantry soldier.

Inside the barracks

There were four platoons per company at that time. I was in the first platoon of the 2nd Battalion of the 186th Infantry Regiment 10th Mountain Division. We were called the Mountain Troops.

Truman's Law went into effect in July 1948, ordering to desegregate the military. The military moves slowly though so we were still segregated during my time there. Our regiment was white and the next regiment of the 185th was black.

Our routine, up in the morning about 6 o'clock, which, for us farm kids, wasn't a big ordeal at all. In fact, we were usually up before that. We made our beds, cleaned ourselves up, got our uniforms on, into formation to head out for breakfast, and back into formation after. Because we were an infantry outfit, we marched...and marched...and marched. And *then* we went to school.

School hours were as simple as watching a movie on fortifications, like how to place a machine gun in a certain position so you could get a field of fire against where the enemy was.

Other school days started with a 5-mile march. We'd load up our gear, carry our weapons and tents, fully loaded, so that we were each carrying about 65 pounds on our backs. And that was just our basic pack. We would occasionally stay out overnight and then come back in the morning. That 5-mile march would turn into 10 miles. Then 15 and so on.

Every Saturday, we had a morning inspection. And when I say inspection, I mean *inspection*. Everything. Our area, clean. Where we slept, tidy. Our footlocker at the foot of the bed, neat as a pin.

It was always a rush on Saturday mornings to try and get everything ready. So, after the first three weeks or so of basic training, we decided to prepare for Saturday's inspection on the Friday night before. We would make our beds, taut as they should be & we'd scrub the floors. It was the end of December 1949, in the dead of winter, in *Kansas*. We, or the ones who had brains anyway, would wrap ourselves up in one of our blankets, place our field poncho on the cold, clean floor, & lay on top of it, just hoping to make it through the night.

Now, when you get a bunch of people from all over the United States, sickness breaks out with all the different viruses and germs. For the most part, everyone gets a cold at some point. And some of us got very sick.

After one of those Friday nights on the floor, I got pneumonia. And one of our soldiers died from the pneumonia that he developed.

When someone got sick, they would take you over to 'sick call' and get a penicillin shot, the magic drug of WWII. You'd get your shot, and then you were back in pronto time. Training had to go on and *you* had to keep on. You marched just the same amount of marching and went to all your classes just like everyone else. I remember watching movies and everyone coughing all around the unheated room. You can imagine how safe we felt, especially me having already had pneumonia. We didn't practice masks, distancing, or anything like that.

Needless to say, they didn't allow us to sleep on the floor on Friday nights anymore. But for the time that we did, our platoon sergeant didn't mandate us to *not* do it, I should say. He went along with it. Anything to make his platoon look good and sharp on a Saturday morning.

Spring season of 1950 and we made it through winter in Kansas. It was now time to take our tests to determine where we would be assigned and where we were going after basic training. These tests determined the appropriate career field where the Army thought we would be most useful. Virgil and I enlisted with assurance that we would get assigned to airborne units of the Army if we could qualify.

Now, at the beginning of basic, I weighed only 114 pounds. I was told I wouldn't be able to go into the airborne unless I weighed in at 125. I guess 114 was too light and they maybe thought I wouldn't descend after jumping out of a plane and they'd never find me, I don't know. So, my mission while at Fort Riley was to put on weight and make it to at least 125 pounds...and I did!

One of the tests at the end of basic was an intelligence battery test. You would have to score at least a '120' on the evaluation to pass. We had to demonstrate aptitudes in a few specific areas that they wanted for that time. One of them, we had to have a high skill in Morse code of all things. It had some situations to work through, like an unwrapped box, only we're seeing it as flat, and we'd be asked, *what form could this be?* There would be 4 or 5 figures to choose from, that sort of thing. Then there was word recognition and vocabulary. I felt I would do pretty good on that portion because I always liked to read.

On one of the days that followed having taken my tests, I was invited into a shaded, *not* well-lit room, I recall, with a very ominous, sinister-looking kind of guy. He played his part perfectly.

There were 6 of us guys who had been able to demonstrate a certain level of aptitude in the different testing parameters on the intelligence battery test. That's how we ended up in that room. We were there to be interviewed.

There we were, the 6 of us, called into this *room* with this *guy*. He talked to us about different sorts of things. And then those that wanted to, went individually into another room with other personnel. This was when they cleared the air with each of us right away.

"Have you got a record?"
"Have you ever been in prison?"
"Have you ever been arrested by the police?"
"Have you ever been in any kind of trouble at school?"
"Any disciplinary problems?"

I was able to honestly answer all of these questions with a firm 'NO.'

They wanted to know where our parents were and where they came from.

They read off their names and things about them and said, "Is this truly your mother & father?"

When all the questions were over and we answered apparently adequately, they just said 'OK.' And we went back to basic training.

This was called a quick surface investigation, which basically meant that while I had been stuffing my face in basic training trying to hit 125 to go airborne after graduation, the FBI was making calls to the Park Rapids Police Department, "Hey, you got an Arthur Burns on your ledger?"

As we got closer to graduation, it was finally posted on the bulletin board where everyone was going. Along with everyone else, I went over many times to check the list, but my name was never posted. Everyone had received their new assignments and were identified on this written list.

With the exception of me…and those 5 other guys.

SO, DO YOU WANNA BE A SPY?

What Could a Life in Military Intelligence Look Like?

The six of us that were unnamed on the specialties assignment list were each called in to the office one by one to be notified that we were, in fact, chosen to attend Basic Military Intelligence School at Carlisle Barracks in Pennsylvania. Should we accept, of course.

Once again, I was questioned, "So, do you wanna be a spy?" I thought, well, it certainly sounded interesting. After all, who doesn't want to be a super sleuth? I accepted.

Soon, I would be off to my assigned specialty of *intelligence*. My friend, Virgil, went off to the airborne.

It was a mixed bag of emotions for me. There was a certain aura attached to the airborne. They had distinguished themselves in WWII and had completed a lot of heroic missions, because at that time, it was still a new branch of the military. Like, parachuting into combat from the sky or the procedure of bringing troops in on a glider, those were new techniques. And to qualify, you had to be at the top of the group physically to even be able to go in. To me, it was both an honor and respectable to be chosen as airborne.

So, I was a little disappointed. Yet, still excited. Besides, there were great stories of spies! Dropping into foreign countries, saving the nation and all of that. I felt good about what it *could* be like.

Some of the other guys from our basic company went on to heavy weapons school or became cooks or even bakers. Others went on to infantry leadership school. A bunch that I knew ended up as engineers, so they went to engineer school where they learned to run graders, caterpillars, how to build roads and fortifications. As I remember, the six of us who interviewed for intelligence, we all accepted.

We graduated from basic infantry training in April 1950, earning our 10th infantry insignia, signified by two crossed rifles. I was now a bonafide soldier.

April 1950

I was granted a two-week leave between my graduation from basic infantry school and the date to report to intelligence school.

Once home, my mom & dad told me that the FBI had gone around Park Rapids talking to different people, asking what I had done, where I worked and whatnot. They even went to see one of my former bosses, Mert Engle, at the Engle's Motors gas station where I worked, who sort of laughed at the FBI's questions and said, "You're barking up the wrong tree, this guy is *OK.*"

Another of my old employers stopped me one day in town and said, "They're around asking about you!" And then the police department, too, "What are they doin' with you in the service?!"

I replied to all of these questions with the stock answer, "Oh, I'm going to intelligence school," or even, "Oh...uhh...I can't tell you what I'm gonna do."

They always told us that if we said anything, we would go to prison for life and all this scary stuff. So, I was kind of brainwashed.

They had spoken to each of my parents while I was away at basic, asking my dad questions about our heritage and how long we had lived where we lived. Really, they just wanted to make sure that our citizenship was real.

While they finished up investigations, I was given my interim clearance. It was time for me to report to Basic Intelligence School.

At this point, I still wasn't aware of what "life in intelligence" could look like, with the job duties or the lifestyle. All I knew was that I just wanted to be a good soldier.

Basic Intelligence School
Carlisle Barracks, Pennsylvania

I quickly learned the reason why I had to have such a high skill for Morse code on that initial battery test. At intelligence school, the first school I went to was High Speed Radio School. Now, I had *never* listened to Morse code. Yet here I was. We had to listen, take down code, and type it. I had to be able to receive the code audibly at 25 words a minute and *send* a minimum of 18 words a minute.

Fortunately, I already knew how to type from a course I took in high school, so I did well on that test & passed the course.

During radio school, I was notified that I received my permanent clearance. This was what they called 'permanent top-secret crypto clearance,' which was the highest clearance you could get at that time in the military. And boy does that kind of clearance make you feel like, 'Mr. Clean!' But also, you couldn't go on to your final classes in intelligence school until you received that permanent clearance. You could even be held back from certain classes while you waited.

Anyway, as it turns out, the agency was specifically looking for radio operators at the time I was selected for basic intelligence school. That was the main reason we got placed in intelligence school and then radio school straight away.

It was around this time that the Korean War broke out, between North & South Korea. I remember we had a formation one morning, where anyone with prior service was asked to step out of formation. Anyone who stepped out went to a separate place where they were screened based on their rank. Reserve officers were sent to officer refresher school. Those who had special training of any kind, say, in demolition, were taken into that special training because there was now a critical need for them. As for me, "OK, you're going to cryptographic school now."

That's right. They sent me off to cryptographic school next. There, I learned how to break into enemy codes and decipher them.

Then, as I was just finishing up with cryptographic school, "OK, you're going to Russian Language School now."

And not just any Russian Language School, but one that had been designed for one particular mission. This mission did not pertain to the Korean War.

We didn't necessarily need to know how to speak the Russian language, but we were there to learn how to *read* Russian text. We learned the Russian alphabet in full, followed by word construction, and then we memorized certain words, titles, names, things like that.

By mid-December 1950, I had successfully completed three different schools at Carlisle Barracks and left for home in Park Rapids before my next assignment.

"So, Artie, what are you doing?"
    "Oh…umm…I can't tell you."

Vint Hill Farms Station, Virginia
Military Intelligence
First Assignment

It was hard to explain to my folks, when I was home, what my life was quickly becoming. Everything was hammered into us, that you can't…say…*anything.*

My dad thought it was super! The Korean War was on, so his main concern was if I would have to go to war. I had to assure him & my mom that I was not. Both my folks were happy about that, that at least I wouldn't be one of the first to be sent to Korea, regardless of what I was doing…and regardless of the fact that I couldn't tell them.

My first duty assignment with the US Army would be as a Strategic Intelligence Analyst at Vint Hill Farms Station in Virginia, right outside of Washington, DC. During WWII, Vint Hill had been a big radio monitoring station, working against the German submarines that were common along the eastern seaboard. Vint Hill tracked them, trying to get a fix on 'em out in the Caribbean and Western Atlantic, along our eastern border. This was a time when there were German u-boat wolf packs, and they would attack our shipping and put their agents ashore and many other *activities.*

From the air, Vint Hill looked like a big farm. It had a big house and some other smaller houses and three huge barns. Of which, two were just full to the brim of Russian material. And I mean, *full.* That's what had been gathered thus far, so for our assignment, that's what they wanted us to go through.

There was lots of trash and lots of insignificant, *"Hello Mom,"* family-type letters. Material could also be addresses of relevant names of important people and other subjects of interest. The industrial stuff, we looked at pretty closely, because that was dialogue between corporations, and therefore, was of great value to somebody, somewhere up the intelligence processing system.

We had to be able to identify and memorize certain keywords so that we could recognize certain important topics if they appeared. When we selected something for further review, we had to type it up, put it on a teletype, and forward it to its respective intelligence desk further up the chain where it was then deeply examined in detailed studies on that subject.

```
                        MISSION:
      To gather enough cryptologic data in order to
            build a baseline of information,
      establishing what's normal and what's abnormal.
```

For the first year and a half, I worked straight day shift, and that's all we did from morning until night. And there was a lot to get through at that time.

Our particular unit was called Army Security Agency (ASA). Although it was considered part of the Army Intelligence Branch, the ASA was separate. This arrangement was changed in 1976 and the entire ASA organization was incorporated into the Army Intelligence Branch.

During my time at Vint Hill, I got to see the newly organized National Security Agency (NSA) take over a larger scope of directing each service in their respective special intelligence activities.

Once we got ahead of the backlog of stored material, I got assigned as the Crew Chief of three other analysts, and then collectively, we worked around the clock. It consisted of four work shifts. An evening shift, a midnight shift, a day shift, and one shift would be on break.

For example, I worked around the clock but at different shifts; 6 evenings, 2-day break, 6 midnights, 3-day break, 6 days, 2-day break. This system worked to support the ongoing operations of that time, which for us, was a live current mission, aside from our regular Russian-related duties.

I accepted that role as Crew Chief because I could both do my job and still do something different every day, to break up the monotony.

In the summer, on those longer 3-day breaks, maybe a bunch of us would get together and whoever had a car would drive us all out to Orange Beach on the Atlantic and we'd spend two days just on the beach. Or, on a really short break, we'd go to Washington, DC and go through some of the big buildings there like the *The Smithsonian.* We spent many weekends in DC because even then it was huge. We'd go up into The Smoky Mountains, maybe camp out for a day and hike the trails. During the apple orchard celebrations, we'd go to the fairs and different events and things.

While on one of our three-day breaks to DC to visit *The Smithsonian* and other sites, an emergency mission came in for a representative from our unit to go to the atomic testing area in the Pacific on Eniwetok Island. I was selected but couldn't be found because I was in DC! An alternate had to be sent in my place. In retrospect, it was probably a blessing as many of the people involved in those testings were exposed to radiation and suffered severe aftereffects.

The United Service Organizations Inc, or USO for short, was also really active then. They hosted dances for us on occasional Saturday nights, where they'd bring in busloads of girls from the DC area and then ya know...we'd dance the night away! The USO provided a pleasant relief to an otherwise dull routine. And you could say, being surrounded by all of these gals in their twenties, some of our guys even met their sweeties and wound up getting married.

I also remember meeting up with two or three of my Park Rapids High School classmates who had joined the Navy. Once their ship pulled into a nearby harbor, they would give a call and we'd meet in DC and celebrate. I remember Jim Collins and Roy Stewart in particular.

Still, the security restrictions of Vint Hill prevented us from saying anything to anybody. In the surrounding areas, if we got stopped and someone asked us where we were from, we would just say, "I'm a soldier at Vint Hill." To which they'd say, "Ohhhhh." Apparently, that was sufficient.

I remember going out in the fall when they roasted the peanuts. There was a big peanut-growing area just south of where we were. They'd have these little stands where they'd bake some peanuts for you. Sometimes they'd soak 'em in salt water, cook 'em so they were kinda salty. And as soon as you stopped at their stand,

"Ohhh, you guys are from the farm?"
"...Yep!"

That's exactly how we explained *it*.

Nonetheless, there was always something that seemed to pop up, always something new to do. Just like when you go away to college, it's more so that you continue to *find* things to do when you're with your friends, right?

I was also communicating with a very special girl, Mary Ellen Johnson, back home at the time. Every so often, I'd receive a letter from her and so I'd write one back. Sometimes we'd plan when we would see each other next as things were getting more serious by then.

For two and a half years, that was more or less my routine, day after day. Very interesting stuff. To me, anyway. Reading lots of material, isolating those subjects, names, and addresses and things, and the functions of daily operations in various parts of the world that were important. I'm sure that kind of thing is all done by computers now…

> **If you can get the DVD, *Three Days of the Condor*, starring Robert Redford, Faye Dunaway, and Cliff Robertson, we did similar things…but without the Hollywood bravado and sleeping around with beautiful women, of course.**

There was a transition period during the tail end of my time at Vint Hill where they started allowing men to either stay on active duty or go home. Which meant there was a small window of opportunity for me to stay on or get out.

The war was officially still on. Drawing to a close but in peace negotiations at the time. This meant that everyone who re-upped their contract was either going to Korea or Japan, almost guaranteed, & very soon at that. By then, the Russians got involved and were at war too, so we were also sending over those with a Russian language background…

February 1953 marked the end of my 2½ year assignment at Vint Hill, the first years of my intelligence career. And to stay on or get out, well, I decided to get out of active duty.

I was not at all afraid of war…
…*I was getting married!*

In March 1953, Mary Ellen Johnson & I got married in Park Rapids. You might say Mary was my reason to get out. Or I could say, Mary was my saving grace, having saved me from that "iffy" time in war, and what could have been, my career & my life, too.

What, get married and then immediately get shipped out? That wouldn't have been a good thing at all. I didn't want to start my marriage that way.

Every one of my friends that re-enlisted at the time left within a month to *replenish* our troops over in Japan and Korea. Some didn't make it home. And because of my Russian background at the time, it was a sure thing I would have gone.

I went back into the Minnesota National Guard & received $300 mustering out pay, as I remember.

Mary: It bought Artie new shoes, a suit for our wedding, and a topcoat!

Mary & Me
Wedding Day

Minnesota National Guard
Hibbing, Minnesota

Right after the wedding and a short honeymoon, we packed up from Park Rapids & headed North to Hibbing, Minnesota where I would be running a Minnesota National Guard Company of the First Battalion, 136th Infantry Regiment, full time.

Up until this point, I had only seen Mary on visitations back home and communicated through letters and short phone calls. Here in Hibbing, Mary & I would have our first apartment together, in the Hibbing Medical Arts Building.

We chose this particular apartment because work for me was only three blocks away. We didn't have a car yet either, so I was able to walk there. It worked out perfectly.

When Mary & I were first married, we didn't have much of anything, not much in the way of furniture at least, and lived 'paycheck to paycheck.' I remember we had an orange wooden crate that stored some of our wedding gifts. With the gifts still inside, we flipped it over, put a new lamp on it so now it was an end table. We had a murphy bed on the wall that we kept down to fill the room.

Fortunately, we met another newlywed couple in the same apartment building who were in the same boat. So, this is what we did. They bought a new dining room set and we bought a chair and sofa. If we had company, they would bring their dining room set over. And if they had company, we'd take over our chair and sofa. Ultimately, when company came for either of us, it looked like we were all doing very well!

Then, we met another couple and decided that we should buy the same kind of TV as each other. This was early in the TV days, so it was quite common for everybody to come over to each other's places to watch TV. It was a brand-new thing! We even had to put antennas on the roof and all that crazy stuff. It was just a part of life then.

And well, with the early age of TV, it was also quite common for one of the TV tubes to blow. They made these snap crackle fizzy sounds, and they blew out *often*. Again, if our friends had company over and a tube went out, they could run over to our place & get the tube they needed. And vice versa if we had company.

Mary: Kind of a neat set up we had there, with our neighbor friends, helping each other out like that. I also had a card table from when I was a dental assistant before Artie & I got married and I still have it. And two suitcases, that if company came, they would sit on those. And that was it.

The National Guard assignment in Hibbing was a whole different assignment this time. It was a headquarters (HQ) of an infantry battalion. I took care of the payroll, supply, maintenance of equipment, and administration of this unit.

One of the administrative sides of my job was to enlist people, trying to fill the quotas of the guard unit. Once I had 'em, I had to get their equipment and all the materials needed for their service and duties. They had to have all their unit equipment ready each weekend or whenever we had our monthly drills and summer field exercises.

Every quarter, we had a weekend drill on top of my regular job duties and that's when we would have our command post exercise (CPX). This is where everybody would be involved in a simulated combat activity or a paper exercise for the battalion (BN) staff only. I liked it because it got me back in intelligence & infantry tactics which had been very interesting and rewarding for me up to this point.

When we had a full scale CPX, all the officers and enlisted members of the BN were involved. The BN commander and his staff & the officers of the four line companies were fully engaged with us.

This is how it went. First, as the BN Sergeant Major (SMAJ), I would contact all the first sergeants of the line companies and establish our communications. Meanwhile, we got our command post center for the battalion staff officers set up and their communications with our four line companies. We had to have our communication system operational between all units of the battalion. All of which had to be done efficiently and correctly.

We would have our maps to work from division, regiment, battalion, and company levels. Then, we'd plot the positions of the 'enemy' and apply the strategies on deployment of our troops in response. All the staff officers would be there, receiving and sending messages as the exercise developed. They would make the decisions and then afterward, the active duty advisors that were at our location would brief everyone, *here's what you did right, here's what you did wrong.* Like, *you didn't move this unit into position on time and if you would have, you would have defeated the enemy faster* or, if you pulled a unit back off the line, *you should have never done that.*

> **Say a platoon goes up on an attack posture and calls over the radio, "I'm getting casualties, I don't know if we can hold." How would you respond, is the question? Would you send him more troops? Or, would you tell them, just hold your position at all costs? Send another force at a different angle, trap, and wipe that whole enemy unit out?**

Everyone played a part in these simulation exercises. Whether they were in supply, intelligence, or other position, they had a part. Even the cooks would be there making the meals. And I believe that's how it still is today.

It was really a good exercise to go through, to keep us in tune with new events, equipment, procedures, and keep us objective about our mission.

Like the Minnesota National Guard, some units got called into active duty for WWII, Korean, Vietnam, and Gulf Wars. In fact, most of the Minnesota National Guard Units, including this one, were called to active duty in WWII and Korea.

In instances like these, you could be activated and deployed within six months, *and* on the front lines *and* in combat shortly thereafter.

Maybe only a single unit would be taken, like the water purification unit, for example, when they were activated independently for the Gulf War. Water was a vital commodity. I mean, it's a necessity anywhere where you have people functioning, right? But that unit, they are specialists in just that.

A water purification unit can take a muddy pothole through a handful of processes and make drinking water out of it.

Things like that were interesting to me so I enjoyed being a part of that.

By 1955, as the Battalion Sergeant Major, I was the senior ranking NCO (non-commissioned officer) in the battalion.

My Korean War service provided me with certain educational opportunities. I decided to take flying lessons! The Government paid 75% and I paid the other 25% towards training costs. At that time, I thought that flying would be fun. So, I decided to give it a try by taking lessons…on the secret. Meaning, Mary did not know.

In hindsight, I must have been trying to keep my dream of being up in the air as alive as I could. I tried really hard to get into it. I trained on *Piper Cubs* and *Tripacer* which were aircrafts of that time. And I got so far into training that I was just about ready to take my final test to get my pilot's license…

…But as sophisticated as these planes were and how neat it *seemed* to be, I actually didn't like it.

It was not a joy to fly.

It was not a means of relaxation.

Nor was it pleasurable.

Maybe I was too serious, but I just…couldn't…relax.

It was just *work*.

I tried even *harder* to ignore all those feelings because I still thought, if I have this license, maybe I can go back on active duty and become a helicopter pilot!

But I kept all this a secret from Mary. Because had she known…. it would have hit the fan.

Mary: And then he left me for two weeks. To go to Guard Camp.

She's right. Once a year, in June, we went down to *'maneuvers'* at Camp Ripley Operations & Training. Maneuvers is where we'd practice everyday military operations. We would either be in combat mode in the field engaged in war simulation operations, or on the firing line in weapons qualification. We would have to qualify with our individual weapons and things like that.

For those two weeks, we established our day-to-day duties and melded them into our, otherwise, everyday job descriptions.

It was when I was on field maneuvers that Mary went into labor with our first child!

I was on maneuvers & Mary was at home in Hibbing. And a friend of ours had to take Mary to the hospital. I had to go to the Commanding Officer of the regiment, who was an older WWII veteran *and* a full Colonel, who said to me, "What's a guy gotta go home for? For a wife to have a *baby?*" To which my Company Commander said, "Well, sir, it is their *first* baby…"

"Oh…okay, well I suppose," the Colonel replied.

The army old timers, in those days, thought that if you really needed a wife, they'd issue you one. Kind of like, *you asked for this son, so why are you asking for privileges now? We didn't tell you to get married!* They let me go home one day early from maneuvers and I got to hold our firstborn, Lori.

For five years, I managed the daily activities of the Battalion HQ, 5 days a week from 8-5. And at one point, I even started working at a gas station in Hibbing at night. It was a good part time job and it paid well but gee whiz, Mary & I were hardly together. I was only home two full days per week. And sure, I slept at home every night, but they were just really long days, every single one of 'em. Every other weekend, I would work a full Saturday. Which meant on those weeks, you can do the math, we only had one full day together. And we were newlyweds.

…But I was "Dad" now. My world had opened up again, just that much more. All of a sudden, I'm realizing…no. I'm responsible for a little bundle of joy now. So, what was I even here for? I wasn't involved in much intelligence work anymore. Was that part of my career already over? Was this even going to be worth it?

Our first apartment was getting kind of small once Lori was born so soon after she was born, we moved into our second apartment. I had just gotten a raise which looked pretty good to us. Our new apartment would be above a garage where the landlord parked his car.

When Lori was a baby, she liked to do this thing. She would sleep all day and was the nicest little kid you ever saw…until the sun went down. And then as soon as we went to bed, she woke up. Oh, that little rascal…

I also remember when the landlord let us put up our TV antennae on the roof. The guys from the guard came and helped me. It was a Saturday and we even worked into the night, turning on all our car lights once it got dark so we could finish installation. They all brought their own drink and snacks, we watched a TV program, & then they all stayed over.

After another year and a half, Mary & I got offered a deal to manage a local motel. This would mean a free apartment for us to live in & a full-time job for Mary with good pay. Sounded like a good idea! So, the three of us moved into the motel.

Mary managed the motel full-time during the day while also caring for baby Lori. When I got home from work every night, I helped out with both the motel and Lori, who was two by now.

Mary and I, but mostly Mary, ran the whole show. Mary was the motel manager but also the unofficial receptionist. Between setting up reservations & checking in all the guests, she was also the fill-in cleaning lady when she had to clean the occasional room. And little Lori helped Mary the best that she could. It was here at this motel that Lori learned how to use a telephone PBX!

"Star Motel? Just a minute, I'll get my mom." - Lori

At this time, I really thought my career & life in intelligence were over. As each day went by, I missed it. Sure, I enjoyed the field training and the weekend training in the guard. It was all very interesting to me. It was productive, too, the simulations especially. But I kept getting that feeling. That feeling when you know that the *job* you're doing isn't going to be your *career*.

Right at the new year of '58, another friend of mine, Bill Gilbo, who was our battalion supply sergeant, was thinking and talking about the same thing as me, going back on active duty. He was at a dead-end job that he didn't like.

By the 1st of February, 1958, we were back on active duty! It still wasn't intelligence, but it was a step in that direction.

That March, I was out on maneuvers again. This time, in Fort Carson, Colorado. I got back one day and on my bed was a note, telling me to call the number that was written. At that time, each company had one phone. We didn't have our own phones in our rooms and we certainly didn't have cell phones. So, first chance I got, I went up to the orderly room and made the call.

When the call was answered, I was shell shocked.

A voice answered, "Hello, this is Folkstead Mortuary?"

After identifying myself, the voice on the other end said, "Oh yes, your aunt wants to talk with you."

Whereupon, my aunt got on the line and said, "Ohhhh, Artie! Your daughter was born today!"

My aunt was over at this 'Folkstead Mortuary' playing bridge with her girlfriends when she called me about Mary. Needless to say, that was a startling moment for me!

Lori was four years old when her new baby sister was born. We named her Jacqueline Rae, after Jacquelyn Kennedy and another friend of ours named Ray Shaw.

Only three days after Jackie was born, I got re-assigned to army intelligence, just like I had wanted. I was to report to Fort Devens, Massachusetts for further assignment.

By that time, I had been off active duty for a total of five years so my higher ups were kind of iffy on if I should go back to 'refresher' school before my reassignment. They talked about my background and the credentials I *did* have, regardless of five years having gone by.

At the same time as their contemplation, an urgent call came in. There was an urgent need in Japan for personnel, preferably with Russian language qualifications. And, well, I had the Russian language digit on my MOS. Right away they said I was going. *Right now?* I thought. Right now, indeed. My

Russian language digit won them over, to say the least, which meant no refresher school for me…I was going to *Japan!*

Obviously, I was caught off guard at first. But soon after, I said, "Well…then…I'd like to put in for concurrent travel," meaning I wanted to take my family with me. The commanding officer said back to me, "Oh, come on, no one gets concurrent travel to Japan!" as if it wasn't even remotely a possibility. After all, things were still too unsettled over in Japan & Korea.

But I insisted.

Over that period of waiting to go to Japan, I tried again & again. "Well…my wife wants me to, so…," and then they started getting a little frustrated with me and by a little I mean a lot. They thought I was delaying and lollygagging against the new assignment when really, I wasn't at all. I wanted to go! But I wanted my family with me.

I put in my requests for concurrent travel regardless of what they kept saying. And as they predicted, the preliminary message came back from our headquarters in Honolulu.

It was refused.

Strangely, though, a second message followed and concurrent travel was approved.

The processing personnel at Fort Devens were shocked to say the least.

Hokkaido, Japan
12th US ASA Field Station
Military Intelligence
First Overseas Assignment

To the best of my knowledge, I would be the first enlisted man to get concurrent travel to Hokkaido, the northern island of Japan, since the hostilities ended in Korea.

The orders were cut! I was authorized for 14 days leave with the family in Park Rapids and 7 days travel time to Seattle, Washington. From Seattle, we would have our car shipped with the little household goods we had and then from there, we would leave by ship, aboard the USS Mann for Tokyo, Japan with further travel to the 12th US ASA Field Station Chitose, Japan.

Mary, Lori, and Jackie had to get passports, immunizations, and all that, to meet the necessary requirements. In the beginning, it seemed at least, that all we did for a while was wait. Wait for the assignment. Wait to find out the day we could get our stuff on board. It was a lot of waiting.

Eventually, we made it to the Seattle Army Base and checked the bulletin board. *Sergeant Arthur Burns and his family report to door so and so.*

The USS Mann was considered one of the larger ships in the Navy troop transport command. We got assigned our own cabin for the family which was really great. The ship left in the morning and we were out of the harbor by evening.

We were headed to Tokyo! And it would be an extra-long time before we saw any of our things again.

There weren't many families on board but there were a few, and with just enough kids for Lori to play with. Lori was so excited, running around with them all over the ship.

Mary: The ship was so big, it would have fit the population of Park Rapids!

Most passengers on the ship were individual soldiers. Many had been to Korea before and others were civilian government families going to Japan. There were three or four military families on the ship that I recall & very highly ranked. But, they were on their way to Southern Japan where families could be accommodated.

The single guys had duties aboard the ship. But if you were traveling with family like I was, that *was* your job. To take care of your own family.

Every morning, we got up and went for breakfast. We checked out the daily newspaper, the *Sea Travel* or something like it, typed on a typewriter down under, & early before anyone had risen for the day. There was always an update posted on our whereabouts at sea.

There were places posted for the women if they wanted to get things washed or iron a few things. It mentioned the activities for the day, which could be a movie or even a dance. There were card games and card tournaments. They had programs where people could get up and sing or a band would play on certain nights. There was a lot of activity on the ship.

One couple had a big challenge. The husband, really a young guy, was severely wounded in war and was now 100% disabled from war wounds. If you're 100% disabled, you get priority travel on all military transportation. The wife wheeled him all around the ship. The two of them stayed in the best cabin on the ship and rightfully so.

During our trip, the two of them taught me how to play bridge. Leading up to our time on the ship, I had watched others play bridge but never played myself. I thought bridge was just for the 'uppity ups,' as if it were only for the 'high society' or something. But throughout our trip, this one player in particular was often unable to play, so the husband & wife couple asked me, on the sidelines, if I could fill in. Or I guess I should just say, I offered.

It was really a remarkably challenging game, and I could see why people played & enjoyed it.

It took 10 or 12 days to get from Seattle to Japan. Our ship's course took the sea-land route so that in case of any real disaster, there was some help relatively close on the main shipping route of the Northern Pacific.

Along the way, we hit a *terrible* storm. They tried to get out of the storm's way as much as possible, but our ship only had so much fuel…so we had to stay on the planned route. Which meant that we, along with the ship, tipped and rolled and rolled some more.

On the main deck, where our cabin was, ropes were tied to everything moveable, even a single chair, and anchored back into the wall so that nothing could start flying around.

People would attempt to get up and then get slammed right back down in their chairs. In fact, I think the water was so rough that the movement put baby Jackie to sleep!

Mary: It was ROUGH. The ship just slung people, right across the room. One guy was sent across the room and landed right in this other man's lap!

One night, we were in bed for just a little while when there was a loud thud against the ship. And to make a ship of that size shudder or shake, well, it's a little concerning. A few of us rushed up and out of our cabins and looked around. And in the wake of the ship, we could see, there was something rolling.

We hit a huge whale! And well, let's just say we don't think he made it. It was a very sad sight…

All in all, we were well taken care of and really quite comfortable. The food was quite good, too. In the navy transport units, they had stewards that took care of the cabins and dining services. A lot of ours were Filipino citizens. (The Philippines was a US territory at the time). They all did a great job.

But it was an experience! And a pleasant one. One that, even with a bad storm & an unfortunate run-in with a whale, we thought was actually pretty cool.

I remember when we entered Tokyo Bay. We could see the mainland off in the distance and over the bow. Finally!

But all of a sudden, a horrible smell. And the closer we got to shore, the ocean's color changed gradually in hue. Someone told us that Tokyo was dumping their garbage into the ocean. And not *just* garbage, but their septic, their *everything*. We heard the old timers on the ship, those who had been to Tokyo or Japan before, sarcastically announce to the rest of us, *"You're in Tokyo Bay!"*

After our arrival in Tokyo, we were processed and then had to wait for our transport to Hokkaido, which took about a week. We stayed in the *Grand One Hotel*. It was the big reception center after WWII and during the Korean War. The hotel was a Japanese super hotel for years, said to be the best building in all of Tokyo.

For that week, yet again in waiting, we made the most of it. After all, we were in Tokyo! *And* I wasn't working yet. We went downtown on the Ginza, the main street in Tokyo & already, it was crazy. They had rickshaws and cabs and the town was all lit up. Just a crazy, showy place.

We stopped at a vendor stand and tried some hot Japanese soup. The Japanese make the most out of everything that comes from the ocean, be it squid, jellyfish, octopus. They dry it in the sun and then recook it. I'm pretty unafraid of most edible offerings, but I had some misgivings on the soup.

Just before we went to Japan, we had watched a Japanese movie, getting ourselves ready for this culture shock. Well, this movie we watched was shot in Kyoto, the 'beauty' spot of all Japan. Beautiful women in kimonos, surrounded by lush gardens. Yet, here we were in Tokyo, in the very center of this huge city, with many smells that were very foreign to us…and very few kimonos. If there were, they were working kimonos. It was an awakening.

Our five days went by quick & before you knew it, we were on the train to Hokkaido, our final destination. We went through the countryside where the railroads were. We looked out our windows and saw the mountains which told us we were on the coastline of Japan.

There were so many marvels, so much fascination. I was excited. *We* were excited. There was this wonder over what was going to happen next. I hoped I had made a good move for my family. There was some anxiety, I suppose a little, like *where were we going to live in this foreign land?* And *will it be an American-type home or something that resembled it, at least?* The "not knowing" and the anticipation felt more like excitement than anything else. It was going to be an adventure, for sure!

By combination of ferry and train, early summer of 1958, we made it to Hokkaido, Japan as a family.

Hokkaido was kind of the poor island of Japan. The area was always a farming community. Beautiful country, land rich in volcanic soil. Even now, they're digging up to 15 feet deep and transplanting the soil to other places.

Before World War II, 'Chitose' had been the location of a large Japanese air base. The Japanese trained pilots there, including those that participated in the strike on Pearl Harbor. After WWII, the US Airforce took over the airbase and continued through the Korean War. At the time of our arrival, the US Airforce was in the process of transferring the base back to the Japanese Airforce.

Remember, I was one of the first to receive concurrent travel. The Korean War just ended so it was still a rarity for military enlisted personnel to be able to bring their family with them to this area of Japan. There wasn't much to offer dependents & housing wasn't really set up for families just yet.

Usually, the guys came over single first, and then once they secured housing on the Japanese economy or when housing became available on base, then they'd send for their families. But the latter took at least a year.

When I reported to my commanding officer, he said, "Come back in a week…just…go take care of your family."

They had to buy themselves time to house us & provide us with resources they didn't yet have, so they encouraged us to take our time settling in. They were kinda worried and concerned because it just didn't happen this way.

It worked for us, though, being given the time to find out where the PX (Post Exchange) was and the hospital and the schools. And really, it was nice that they wanted to make sure my family was processed correctly and that everything was okay with them.

In preparation for more families, they started to modify the PX and commissary and convert the otherwise traditional single-soldier Quonset huts into living quarters for families like ours. A Quonset hut was an emergency [quickly] constructed home for the air force during the Korean War. Traditionally, four guys would live in one quad and in the center was one communal bathroom to be shared by all four guys.

But just for us, they reconfigured this Quonset hut for a family. They made us a couple of bedrooms, a kitchen, bathroom, and living area using the same, usual quad configuration. And this would be our house. It was quite spacious, as far as I remember, and warm.

Meanwhile, the Army was building a brand-new housing area called 'Chitose 2' about 5 miles from the air force base. The ASA troops had already moved to that area because they wanted to get our operational people to where they could work in a classified, secured area. They were trying to get us out of the city because it was growing rapidly around the air base. They called that a boom town in those days. Just like if you go to any American base now, it's always situated next to the boomtown, or the fastest growing city in that area.

As for school for the dependent children, frantically they contracted with any former teachers that were on base. Due to the state of emergency, some of the soldiers' wives that had been teachers in the states were given authorization, regardless of their credentials, to teach that first year that we were there.

Everything was new and we were part of the experimentation along the way.

The only paved road in Chitose
Downtown Chitose

Downtown Chitose in the winter

It took about a month before I was really put to work. After that short time of adjustment, I was assigned and reported into operations.

We lived in our temporary quarters for about three months and then the family quarters in Chitose 2 were completed. Almost immediately, they offered us a two-bedroom unit.

The new set up was of duplexes, about 5 miles down a dirt road. And I mean, really a dirt road. This was okay during the dry season, but when it rained, it was something else. The commissary, hospital, school, and everything else was over at the old air force base. So, all of our shopping had to be done down the ol' dirt road.

Entrance to 12th US ASA Field Station & the ol' dirt road

One of the first things they told us, once we were really settled, *get a maid! She can help you...*which began our life and relationship with Takako Kanayama.

We found Takako through the employment office where they had listings of girls looking to do maid services in the homes of US service families. She was native to the area.

When we first met Takako, we didn't really know what to think. Why, there was just this young woman in our house all the time now! At first, she was timid and shy, trying to please all the time. But then, she just took over. She knew just what to do.

She would walk right in and immediately check on the kids. Get them dressed, fed, and off to wherever they needed to be or to whatever she wanted to do to entertain them for that day. And my shoes were always shined.

She loved Mary and she loved little Lori, and then to have baby Jackie, why, she loved us all unconditionally.

There was also a socio-economic element to hiring someone like Takako. In a sense, it was our way of supporting the recovery of the Japanese economy after war.

Japanese locals, like Takako, were curious about us. We were a representation of 'American family life.' You could show one of the locals a picture of your home in the states & they *wanted* to see that. All they had for reference was American movies, most of which were cowboys and Indians shooting at each other. And for a long time, they believed that's how it actually was!

Takako was young & curious about the American culture. She spoke very good English & wanted to learn American customs, or our American 'ways' so to speak. She was always willing to learn how to cook & how Mary did things in the kitchen.

For the kids, Takako was a big role model. She was their teacher, their surrogate Mother, their auntie, their older sister. She taught them fun and common American words in Japanese. She taught Mary how to cook rice the really good Japanese way. And aside from helping in whatever manner she could, she especially helped us with any cultural questions we had.

And then she invited us over to her own home. Which was a rare thing. It was just her with her mother, & our family of four. A cultural event for us all.

Here was this house that they lived in, with only one bedroom. That one bedroom was for the *papasan,* the father, who could deny his wife the privilege to sleep in that bedroom. The rest of the family members had bed mats that they put on the floor. They took them out of these little storage areas that were specifically for holding bed mats and put them down on the same spaces that are used for everything else during the day. Where they lay their bed mat at night is the same place where the table comes out, tipped against the wall, where they would eat. This is why no one wears their shoes inside a Japanese home. Because the floor, it's their table, their bed, their everything. Therefore, it is kept clean, *very* clean.

Inside a Japanese home, you wear these soft house slipper type socks called *tabi.* And once you know the Japanese customs and why, it makes sense. That you can live in the smallest place possible, yet sanitary, clean, and most importantly, livable.

You don't fully appreciate other cultures until you see and live like that.

So…it would make sense that the only thing Takako demanded in our house was no shoes. *Don't dirty the floor.* She was always cleaning our floor, never wanted us to do it. Sometimes I would forget to take my shoes off when I would come home and then I would look up and see her staring at me. She'd polish our floor by wearing a pair of my wool socks and sliding around the house.

Mary: I would lay down a sweater and Takako would pick it up, fold it, and put it away before I could think twice. She was my friend. And we were the same age.

We never told Takako or asked her to do really anything. She knew what to do. We didn't have to tell her, nor did we have to correct her. And she was just a sweetheart. Like a permanent babysitter & much more than a maid.

As for the kids, they loved Takako and she loved them back. We even took Takako on our family vacations. She truly became part of our family unit, & for that, we feel forever fortunate.

In front of one of the Quonset huts we lived in
Takako holding Baby Jackie,
& Lori

By the time I got situated at work, they really needed me, so it was a good time to be there. I felt that it would be a really good job and I knew I was qualified for it. My normal hours of work were 8 to 5, five days a week.

However, due to my specific duties, I was on call 24/7. Whenever the phone would ring at home, it usually meant an urgent summons.

I got to be home just about every other night except *those* nights, when I would get called in, which sometimes meant I would unknowingly be away for prolonged periods of time. Sometimes I'd sleep in the room where my office was. I'd turn off the lights and my friend & I would take turns between taking a nap and working. The opposite guy would stay right on position.

Mary: I never knew where he went. I never knew anything or what was going on. In fact, every single time he left, in Japan, we had an earthquake! I could be lying in bed, when the windows would rattle, and my bed would whoosh across the room. By the time I got out of bed and got my children out of bed and under the doorway, it would be over. Every time Artie left.

What Mary didn't know and could never know, was that any time I was ever gone, in Japan, I was usually only about 2 miles away from home.

Mary: I never knew that. I absolutely never knew that until right this moment.

Delirium

It was winter, a big storm was on its way. I remember the snow had just began to fall. This one morning, practically delirious, I knew I was *really* sick, with what felt like an appendix attack. But I truly believed that I couldn't go to sick call until I was fully dressed and in uniform.

When I got to sick call, there was a whole line of other sick guys. With the exchange of only a glance, and only a few of my words, I was taken right in to see the doctor.

The doctor took one look at me and put me on a big table, poked around, and told me that because they didn't have the right facilities for what I needed, that I needed to be transported to Tokyo.

But due to the storm, all air flights in and out were cancelled.

"Well, we gotta operate on him here," they said.

And that they did.

The first thing they did was give me a spinal block...or they tried to give me a spinal block. My legs were the only part that numbed. It didn't work anywhere else.

I remember him poking me in the abdomen next with a needle because *oh MANNN!* Let's just say I gave way to the fact that I was *not* ready for an appendectomy.

I remember as they wheeled me into the operating room, I looked up and saw Japanese painters right above me, just painting away and having a good laugh at me down below.

"Too late. Gotta do it NOW," someone said.

Onto the table, two big dudes held me down. They were practically sitting on me. With zero pain blockers and the anesthesia not working, right on the operating table, *completely conscious*, they cut out my appendix.

They scraped, tied things off, and cleaned up with alcohol. Like a *fiery hot iron sizzling my insides* -- is the only way I can put it.

I stayed at the hospital on base. And even though she was pregnant with our third child (that's right), Mary stayed by my side while Takako was at home with the kids. When the doctor came by, I said, "Doc, I'm not a baby, but that really hurt." He assured me that he understood & that I did a great job.

Before I dated Mary, I gotta say that I had been going with another girl. She was one of the first blondes I ever dated, maybe even the only one. A few months after I went into the Army, she started dating this super jock, nice guy I'm told, and Mary knew him. He was a classmate of hers in high school. His name was Guy Inabinet.

So here I am in the recovery room all the way in Chitose, Hokkaido, Japan and Mary walked in. But wait….is that Guy Inabinet with her?! Here I am, thinking I must be in a pure state of delirium, that this guy's followin' me around! Apparently, he was a helicopter pilot & flew in from one of our ships to get some things. But right there, he picked Mary up, pregnant as can be, spun her around, and…

Mary: …Here I am walking in…with this guy that took his girlfriend away from him years earlier!

And there I was, looking like a truck had run over me. A hard to believe story!

The Other Mrs. Burns

There was another Sgt. Burns and family that were in Chitose at the same time we were there. His first name was <u>A</u>ndrew, mine <u>A</u>rthur. His wife's first name was <u>M</u>oe, and my wife, <u>M</u>ary. They had 2 daughters about the same age as our two daughters. Both our wives, Mary & Moe, were pregnant and were scheduled to give birth to their third at approximately the same time, August 1959.

A little while after our son, Tommy, was born, the other Mrs. Burns next ·door went into labor. Her husband, the other Sgt. Burns, was out hunting somewhere, so he was nowhere to be found. Therefore, *I* had to take her to the hospital.

We got to the hospital, they took her in, & I waited in the waiting room. After a while,

"Sergeant Burns? Get in here!" called the nurse. Innocently, I followed her orders.

Before my eyes, the other Mrs. Burns was all prepped in stirrups, legs up and open, ready to give birth to this baby. In fact, the baby's head was just starting to show and I had never seen anything like it in my whole life, nor ever after.

"Well, hold her hand!" The nurse exclaimed to me.

Again, I obeyed and began to hold her hand. Mrs. Burns even looked up at me with a strange look and I, too, looked down at her in the same kind of, *oh heavens* moment.

The baby is born! And the nurse asked,

"Well, what are you going to name him?" And looked at *me!*

So, I said,

"Well, I don't know…she's not my wife and the baby's not mine!"

Yelling and hollering broke out.

89

Now I had to explain who I was.

I assured them that although I was Sgt. A. Burns and although my name tag spelled it out, I was the *other* Sgt. Burns. The one next door to them and that her husband was out hunting!

"She needed someone to take her to the hospital and it wasn't the time to discuss who I was!"

That's how I left it.

A few days later...

Wouldn't you know, it happened again, within days, with another neighbor. This time, I stayed in the waiting room. Their names were Bill and Evelyn Falk, and again, they had their first son.

Word got out around base, *if you want a boy, have Art Burns transport your wife to the hospital.*

However, I don't *ever* want to see another birth. I wasn't trained for that. Actually, I don't think any guy is supposed to see that kind of thing. People tell me things have changed and most husbands go into the labor room with their wives, but I'm not so sure I'm sold on that one.

In Japan, it's a great celebration to have a son. They fly a big carp flag, *koinobori*, outside your home, which signifies that you have a new baby boy. Before you knew it, we had three koinobori flying from each of our family quarters, with three small sons in our little corridor, all born within a month of each other.

Mary with Baby Tommy, Me with Lori, Takako with Jackie

The family, Christmas time in Japan

An American Embarrassment

One night, Mary went to a Japanese origami-type flower arranging class with some of her lady friends. The military women, there in Hokkaido at least, were all free to socialize, as most had a maid at home to babysit. This origami class was one of the offered 'life learning classes,' as we called them.

On this night, I stayed home and got to watch the kids. One of my friends from the military base police department (MP) stopped by to see me for a while.

Just before Mary came home, there we were, just hanging out, when I received a call from the officer of the day, who asked me,

"Were you up town? And did you run over somebody?"

I held the phone to my ear as I watched my three kids run about to and fro & assured him I was not uptown…nor did I run someone over. Unfortunate for me, my kids all under the age of 5 could not be legal witnesses. But perhaps my friend who stopped by, could. I mentioned who I was with & ensured him where I was in that moment, and that was the end of our phone call. Resolved, I thought. Wrong!

The next day, I'm home & get a knock on the door. I opened the door to see a disgruntled officer, "Sergeant Burns, your license plate has been given as the car that hit and injured this guy up town last night. You need to come with me."

Mary: I had to say goodbye to him, not knowing when he would return!

I was being charged with a hit and run. By some guy up in the bar district, a place I had *never* been.

Off I went, down to the provost marshal. They set up a line of suspects and I had to get in the line with them. The two that were 'run over' couldn't identify anyone from the line *and* they looked right past me.

In the next day's issue of the *Shin-bu Press,* a Japanese communist newspaper in Sapporo, the provincial capital of Hokkaido, was published the following article:

Arthur Burns, an American war monger,
ran over two of our poor indigenous workers in Chitose,
half killed them,
& then ran off from the scene of the accident
and to his American compound.

Soon after, further investigations concluded that it was, in fact, someone else. Not me. And in the police report that followed, they clarified the mix up.

The guy who turned in the claim had juxtaposed the last two numbers of the license plate. So instead of 87, it was 78. Which, in the end, matched my plate.

"Two drunk Japanese walk out of a bar," I suppose you could say…and walked out into the street and right into someone's car! Half in the bag!

One of the guys even wrapped himself in cloth, faking a hurt neck. It turned out that both the two guys were imposters and later arrested.

But that's not all.

As for me, I had to go in front of the provincial judge in Sapporo and apologize for me having caused this incident.

See, the Japanese philosophy is an interesting one, one that I understand now. It was the mere fact that *none of this would have happened if I wasn't there.* If I wouldn't have been in that country, I would have never been called because there wouldn't have been a license plate with those numbers with which to get juxtaposed and confused.

And even though they didn't catch the right guy in the first place, *I, Arthur Burns, caused the Japanese government and people of Japan a great embarrassment,* and was requested to apologize to the judge for all the trouble I had caused.

Therefore, I stood before the judge and begged for forgiveness.

"Alright, Mr. Burns.
I see your heart is in the right place.
You can go."

In the next *Shin-bu Press,* it read,

Sergeant Arthur Lee Burns,
warmonger,
was not the right person...

And then they named the correct American soldier driving the correct car.

Fortunately for that guy, he was with some other Japanese people that night who refuted and corrected the whole story. No further apologies from the *Shin-bu Press.*

A sobering and kind of scary experience.

My three-year assignment in Japan was coming to an end. So somewhere within the last six months, knowing that we were soon leaving, I found Takako a job to replace the position she had with us. Her new job, upon our departure, was with an international telephone company. She was happy and we were happy for her, too.

One item of interest I would like you, my reader, to contemplate.

During the time of my assignment to Japan, the height of the space race was on, between the USSR and the USA. The Soviets were ahead of us during this period, as evidenced by their successful launch of Sputnik 1 on October 4, 1957. Its orbital track went right over Hokkaido. Every 88 minutes, we could stand in our yards and watch it pass over us.

During this same time, the Soviet impact area for their long-range missile program was on the Kamchatka Peninsula, just north of Hokkaido. It is very interesting that I and many others were stationed on this remote island at that period of time.

Don't you agree?

Sergeant Arthur Lee Burns
Class A Uniform

Life in Japan
Through Lori's Lens

My years in Japan are great memories. I was young, only four years old when we moved and then left at the end of kindergarten. I don't know how my mom & dad did it.

Dad would get in the car, and just start driving on the other side of the road like the Japanese, adapting right into the whole lifestyle and everything it entailed.

I remember sitting out on the sidewalk in front of our little Quonset hut watching the gardeners. Nothing was more interesting to me than to watch them eat their lunch.

I would watch this one man, day after day. He'd open his bento box and the things I would see in there were just fascinating to me. Like the little patch of sardines or carefully curated fish pieces. I would just sit there and watch him eat like it was the most wondrous sight in the world.

And there were loads of kids, just like us, in Hokkaido. We would zoom in and out of every house, just a bunch of kids flying around.

I remember seeing and even feeling the different dynamics from one home to the next. One of my friends had very strict parents. In fact, in most of the military homes I went, things were run tight. I really hated that. But I'm sure the values of being a soldier got passed on to their family units and how the home was run.

My home, though, was such a safe place. There was never a moment I didn't look forward to being at home. It didn't matter that my dad was in the military off doing military things. Home was safe.

In kindergarten, we visited a Japanese school. Mom came as a chaperone. We learned quickly about things we'd have to do in a typical Japanese school. *No shoes allowed!* Just like Dad said. That was the first rule. Those special socks, *tabi*, were like little gloves for your feet. There was a separation from the big toe to the rest of the toes and foot so you could wear them with wooden platform shoes & you could go through puddles without getting your feet wet. Every child had a pair when they went to school, so all they had to do was kick off their shoes in the doorway and wear the tabi indoors.

Hokkaido is cold, like, really cold, with winds coming from Siberia. The kids wore lots of layers & had puffy jackets. Theirs were very thick, though, not lightweight like the ones we have now. There was always a stove for burning a little bit of coal because there was no central heat. We'd huddle to keep warm and teach each other songs.

Although it was Dad's military career that took us to Japan, Takako was our *connection* to Japan. What I remember most was the bond we had with her.

Mom was busy and then Mom was pregnant. She just had my little sister, Jackie, and then was pregnant with my brother, Tommy. She needed the extra help and boy, did Takako step in.

Takako learned from our family just as much as we would learn from her. I would wait for her every morning to walk through the door. I would not get dressed into my day clothes until she was there. I would sit in my pajamas in the chair and wait. She may have not appreciated that at the time but that is what I did.

I was Takako's shadow. I got an inside look at everything she did and everywhere we went, dragging me along from place to place. She would take me to ballet lessons in the village and she would take me to her hairdresser. The Japanese ladies would walk up to me and want to feel my crazy curly blonde hair that was oh so different from theirs. They'd touch it and play with it and were just in pure wonder over this hair of mine.

Back in those days, they dug ditches in the back alleys filled with running water where you would go and relieve yourselves. Not like the outhouses that we have here. But this was post-war Japan and life for the Japanese was tough in Hokkaido. When we were out and about, Takako expected me to squat in that ditch…and expose myself to the world! I refused. I know she would get so frustrated with me, *girls have to go potty, just do it!* Yet, I was horrified.

I remember going to Takako's house, seeing rice paper sliding doorways that kept the heat contained between rooms.

Takako *wanted* to take me to places, just as much as I wanted to go with her.

Takako Kanayama was the best. I love her to this day.

Me, Lori, with two friends in Japan

It would be a real experience getting out of Japan.

We left Chitose as a family & flew down to Tokyo to the processing center. Before it was our time to leave, most flights were by *Electras,* which were large aircrafts of that time. But they were having lots of trouble with them. Some were even crash landing.

When it was our time to leave, they had just grounded all *Electras.* So, they had to split up all personnel that were scheduled to depart, into two different aircrafts.

The bulk of us went on what's called a *Super Constellation.* They didn't have a very long fuel range or baggage compartment so when they were loading us up, our family of five, they said to us,

"We're gonna be making frequent stops to fuel up so you can have one suitcase...for all five of you."

We had *five* suitcases! Mary & I with our bare essentials, but then Tommy was a baby, so his suitcase had his diapers and numerous changes of clothes for the trip.

Yet here we were now. Flying on a *super constellation* with *one* suitcase.

Tommy's.

On this trip, Tommy, had a *terrible* problem. He drank say, two ounces of milk and then threw up four.

First, we flew into Wake Island, one of the most isolated islands I have ever been to. Then, we went to the Midway Islands and stayed overnight. They fed us there and we got a chance to refresh ourselves but still no suitcases.

After our stop in Midway, we then went to Guam and at last, arrived in Honolulu. At this point, Tommy had burped on just about everybody he encountered so maybe you can imagine how we all smelt.

We got to see all of these new places on this trip which was interesting, but with each stop, we were a little grimier than the last.

Hint: sour milk.

It was quite the trip.

By the time we made it to San Francisco, it was a 3-day trip in total, which should have been one.

I had taken all our money from our bank in Chitose and put it in a bank note for our bank in Park Rapids. Which meant that when we got to California, we couldn't access any of our cash because it was exclusively for the bank in Park Rapids to cash.

This certainly made it difficult to buy travel tickets for a family of five to get home all the way from California to Minnesota!

We had to have our parents wire us money and then wait for it to come through. Luckily, Mary's dad sent us enough to get home. We opted for the cheapest route which was the train. But really, we also thought it could be pretty cool for us to take a family train ride, practically across half of the United States. Once again, we chose to make the most of it.

But I'll tell ya, I felt like a real pauper coming home.

The train took us from San Francisco and cut right through Reno on the main street. We had a sleeper which could have been comfortable but the air conditioning in our train car quit by the time we were only *half*-way home. I should mention for timeline purposes…this was the hottest summer on record. As we traveled through the State of Iowa, the train personnel let us take turns sitting in the other air-conditioned cars. The second half of this trip was a challenge to say the least.

One strange occurrence happened before we left Japan. My new assignment was to Fort Devens, Massachusetts. Unbeknownst to me, the 'other Sgt. Burns' received orders to Fort Meade, Maryland. Well, he didn't want to go to Fort Meade, so he convinced the personnel department at Chitose that there was a mistake in our assignments because of our names. Successfully, he convinced them that the orders were mixed up. So, they changed our orders, assigning him to Fort Devens and me to Fort Meade.

He left for Fort Devens with his family in the spring of 1961. Later that summer, I left my family in Park Rapids while I found housing at Fort Meade. When I reported for duty in Fort Meade, they informed me that there was no mistake in the original assignment, and I was supposed to be in Fort Devens while the other Sgt. Burns was supposed to be here where I was in Fort Meade! By then, the other Sgt. Burns had now convinced his superiors of the mix-up and found a permanent position for himself at Fort Devens! I, on the other hand, was left to wait it out to get new orders for transfer to Fort Devens.

It would end up being three months of wasted time.

Finally, I received my new orders to my correct location all along, Fort Devens, Massachusetts. I went alone to find living quarters before bringing the family. They stayed back in Park Rapids as Lori was in school.

Unfortunately, I found no quarters for the family on base but after a long search, I found an adequate apartment in Leominster, Massachusetts, about 9 miles west of Fort Devens. I took another leave, went to Park Rapids, and brought my family to our new place in Leominster. This would be the 2nd change of schools for Lori.

Four months after our arrival, the construction of new quarters on base was finished and we moved in. This was now the 3rd change of schools for little Lori, in just one year!

From the summer of 1962, things stabilized and got much better.

Fort Devens, Massachusetts
Intelligence Instruction

Fort Devens, Massachusetts & another new role. This time, I was going to be an instructor! At the Intelligence School in Fort Devens.

First, I had to go to what was called 'charm school.' Here, they taught the proper way to present oneself. After all, I was going to be the head of a classroom of students.

A funny one was to, *never turn your back on the students when you wipe the chalkboard,* especially in a back-and-forth horizontal fashion because then you get to wagging your tail and the class might laugh at you. Up and down was the way to go so that you stayed tall and straight. Ultimately, though, I learned to always be looking at the students.

At charm school, we were highly judged on both our physical appearance and the way we wore our uniform. We had to wear any medals earned and our name tag, & everything was judged. The instructors in our audience would then rate us after we were done.

We learned the different types of instruction, be it lecture, demonstration, or practical exercises. Each had a different style of instruction which I thought was interesting. We learned how to give a presentation, how to give a full verbal lecture, and then how we would teach a class when showing a film. What was important about our instruction was that the students didn't have four years of classes or subjects to graduate. At most, these were six to nine-month highly concentrated courses.

Toward the end of charm school, those who were going to be instructors had to then give a class to the current instructors. Mary trimmed my hair before the big day, and whenever else it was time for a haircut really, as there were no barbers in the immediate vicinity.

My students were primarily Army, some Marines, a few Air Force, and a few from the Navy. Each course had a name and a specific goal with about 40 students per class. The first course I taught was *Crescendo*.

We started with the basics. We presented this program to the class, that with the information fed to the students, this research endeavor grew and grew to become, with the right details, 'the big picture,' hence the name *crescendo*. The full picture, I should say. Then, after taking a glance and reviewing the full picture, we would then implement *Klein*.

Klein was a great representation of how intelligence processing works. *Klein*, meaning 'small,' meant taking one section of the full picture and zeroing in on one specific section. Breaking apart bit by bit this one area or this one piece of information, taking it one step further, to hopefully gain even more information on that one thing.

Just like intelligence, the students would report a general perspective of a situation in its entirety. The next step would be to select what needs to be explored further and exploring it so far that there would be no piece of valuable information left uncovered.

MISSION:
With these courses, we were teaching both the theories and procedures that were fundamental to everything that they would ever do. To get intelligence out of information gathered. Students wouldn't have to break any security codes because we wouldn't mention any specific nation or target.

Many of these theories and applications had been taken directly from WWII, learned in successes in intelligence collection and processing. As their instructor, I emphasized that when they left training, things could be a lot different than the way we were simulating them to be. But when they did go on to their respective assignments & apply the principles and procedures taught in these courses, that they *would* produce results, and just about always successfully.

During those classes, our students would usually receive their final security clearance, graduate, and after, be assigned to their units or service component. It was also common that many of our enlisted men and officers, after some field assignments, of course, would get cycled back to be instructors, making them more effective because of their own experiences. They'd get the chance to learn from their own mistakes & how to handle problems more effectively, making them better teachers in the end.

In WWII, the US Navy broke the code of the Japanese Navy. Before the attack on Midway by the Japanese, US intelligence approved for our Navy to report in plain text that they were 'having trouble with their water purification' which was a hoax. As hoped for, the Japanese reported back to Tokyo that 'Midway' was having trouble with their water purification. But they didn't say 'Midway,' they said 'QM,' as in 'QM is having trouble with their water purification.' So, there was a continuity error of sorts. We didn't know what their codes were at that time or what 'QM' meant. But with those two pieces of dialogue, we knew. We knew an attack was coming.

This experience taught us all that sometimes in strange places comes something that is invaluable to the rest of the intelligence gathering. The huge naval 'Battle of Midway' that followed on June 4, 1942, resulted in an overwhelming victory for US forces, the first success for the US against the Japanese since Pearl Harbor. Get the DVD, *Midway,* for more details.

At the end of each of my classes, I kept feeling like something in the instruction was missing. Things were changing fast and evolving rapidly at that time. But by this time, there had been a particular process implanted in me. *You can always explore even further than what you think.*

That was just it. In my spare time, I started crafting a lesson plan on that very one-step-further concept, and just how to proceed down that route. How it could really expose the enemy, and most importantly, how you can then capitalize on the enemy's self-assurance.

For developing that lesson plan, I received The Army Commendation Medal. After I implemented the lesson plan into the classroom, I was promoted to Master Sergeant. It would remain part of instruction for years after.

The Army Commendation Medal

The Vietnam War was building up toward the end of my instructor assignment in 1964. As instructors, we had to start developing more tactical, *quick* crash courses. Before I knew it, the mission became to train the intelligence units going to Vietnam in the shortest amount of teaching time possible.

When it came time for my next assignment, I had two options. Germany...or Vietnam. And those were my *only* options...

US ASA
European Central Processing Center Headquarters
Frankfurt, Germany

...Thank God they sent me to Germany.
Because of my language background, it was a good fit.

After having moved quite a few times by now with the family, within the United States and once outside the US to Japan, being re-assigned to various places was just becoming a part of our life. I had learned so far that so long as I held a meaningful role within the field of intelligence, I would be content.

What helped tremendously though was that Mary was happy wherever we went. Although concerned at times, she accepted our relocations and accepted them well. Mary developed her systems and means of getting through. She found a balance between being self-sufficient, independent, and my companion. I started to become very aware and grateful to have chosen such a partner in life, to be able to pursue a career in the military that both suited me and was supported by my wife.

All this to say that when we got selected to go to Germany, Mary joined in my excitement. We were lucky to get concurrent travel again so the family would get to go with me. Lori, the oldest, was excited to go, too. In fact, we were happy about the whole circumstance exactly as it was. Eager to experience another new culture and confident that since we had done it before, we could do it again. I think that comes with both knowledge & experience of travel and foreign countries.

As we knew it, my career was actively teaching us, as a family, our children especially, how to adapt to and embrace change.

Right from the beginning, Europe proved to be quite different than Asia. The only other foreign living experience we had up to this point was Japan, which was very different to say the least. The European culture and their traditions were similar to ours, so for us, easier to assimilate.

We processed out of the US at Fort Hamilton, New York City. We departed New York harbor aboard another converted troop ship, this time, the *USS Rose.*

It was another large ship, like the one we took to Japan. As we left the harbor, we passed the Statue of Liberty and Ellis Island, where most of the immigrants from Europe were processed upon arrival to the US in the 1800s.

It was an uneventful yet beautiful trip across the Atlantic. We went through the English Channel & passed by the White Cliffs of Dover. I was impressed by their size & beauty. We docked in Bremerhaven, Germany, located in the Baltic Sea, & took a train from the port to Frankfurt, our final destination. On the train, we had a separate family compartment for four that was really comfortable. But...big sister Lori had her *own* compartment.

Lori: I had the best sleep of my life on that train, with the rhythmic sound of the train wheels on the track. I slept so hard that I didn't hear my parents pounding on the door to wake me. They eventually had to get the conductor to open the door!

When we got to Frankfurt, we didn't have government housing, so we lived on the economy while trying to find housing in the German community. It just so happened that a friend of mine who had served with me in Japan was also assigned to Germany at the same time I was. Lucky for me, he had been stationed in Germany before, so not only did he have a good grasp of the language, but he had also gone through the 'house hunting' process before.

One of the first few days we were there, he took me around town to help me find a place for my family. Regardless of what he knew though, we did still struggle to find a place at first.

He took me to another nearby town, Ober Erlenbach, and walked right into the one and only bar. Bars in this area were like our US community centers, where people meet and hang out together. In fact, a bar in Germany is called a 'gasthaus,' which translates to 'restaurant/guest house' in English.

With the little German that he knew, he asked if anyone knew of a place to rent for a family. They talked and talked for a little while and then one of the Germans told us about this lady, Frau Kunz, who was renting apartments nearby and that she was not very far from where we were.

We found Kathe Kunz, owner of 3 housing units. Kathe's husband was a painter by occupation, which they called a *malermeister*, 'maler' meaning paint and 'meister' meaning master. He fought in WWII in the German Army. He had passed by the time we met Kathe, who was then not only a widow, but lonely of heart.

Kathe lived with her three sons, Walter, Gerhardt, & Helmut. Helmut was the youngest of the three and spoke the best English. Therefore, when we came to ask about a vacancy, she told us that before we could discuss any further, we would have to wait for Helmut to get home so that he could translate.

Helmut arrived and Kathe then agreed to rent to us while we waited for government housing. Our family would be the second of many American families that she would rent to.

We took occupancy of an upstairs unit while Kathe and the boys were next door in the main house. Kathe was just as delighted to have us as we were to be her guests. And I remember everything being so funny to her! Probably because she got to watch an American family of five try to adapt to German living arrangements. Which also meant that Kathe would soon get used to what an American family would want or need.

We Americans were accustomed to washing machines and dryers and other appliances that European families did not usually have. In Germany, we were putting coal into this little furnace. We would then run water through an attached boiler, heating it up so we could take our baths. Everyone had to try to take their baths while the water was still warm. We also didn't have any experience in fueling with coal to begin with so that was a tricky job to learn, and we never mastered it very well throughout our time there.

Americans also use an abundance of water whereas Germany is more resourceful. They recycled the water. Two or three members of the family would use the same wash water and then 'recycle' the water when done by throwing it in the garden. These are just a few examples of 'living on the economy.'

Being neighbors with the Kunzs' worked to Jackie & Lori's advantage. In Japan, they learned Japanese from Takako. Now in Germany, they were able to pick up some German from the Kunzs.' And just like Takako got to learn more English, the Kunzs' had the opportunity to fine tune theirs, too. It was another perfect scenario.

Our rental apartment in Ober Erlenbach

Once we were settled at the Kunzs' house, it was time to get to work. This time I would serve in a supervisory role for the US Army Security Agency at the European Central Processing Center Headquarters. I would supervise a total of 40 data processing personnel & analysts.

The intelligence mission was similar, as it always had been, just on a different target, which I can't disclose. For this target, as with many others previously, we extracted data & prepared reports. There was even an element of 'techy' expertise needed when they started implementing new computer systems.

The nature of my job though was safe, and more or less without threat. I was never in harm's way…that I know of.

Eventually, we did receive military housing and had to move out of the Kunz apartment.

Our new development was right outside of Frankfurt in a small town called Fischstein. We moved onto the second floor of a three-story complex. Each apartment was nice and had its own balcony. Above us was another family that we were able to grow quite close to.

One Christmas Eve, I asked the neighbor friends above us, "Would one of you go out on your balcony and drop this bag down on our balcony, 'get it swingin' and throw it down?" We talked about the timing and logistics to make sure we got it right. "Yeah! Of course!" they said. I explained what we wanted to do and everything and they were more than excited to help me and 'Santa.'

A little while later, I said, "Hey kids, I think Santa is going to be here soon, I'm hearing bells outside——," then *BOOM!* This bag fell down onto our balcony. We all looked around and the kids said, "I don't see him!"

All of us ran to the big picture window in the front, and I said, "There he goes!" Jackie shrieked, "I see him!!" And maybe little Jackie did, I don't know. But the kids, man, they were stunned beyond belief.

Jackie: The stage had been set! Dad would try to top that every year after.

Lori: Yeah. It was probably the most impressive sight for a child. Since I knew there wasn't a real Santa by this time, it had me thinking twice!

The highlight of being in Germany with my family, though, was time spent at the Frau Kunz apartment complex. We all became so incredibly close to each of them that they invited us over to their home for Christmas, even after we moved out of their complex & into government housing.

That Christmas, the Kunz family decorated their place in beautiful Christmas lights and had real lighted candles on their Christmas tree, which were lit only for a short time during the festivities. They even had sparklers on their tree! And then, there was a knock on the door.

Lori: It was Christkindl. She's the appointed town angel that comes to your house and brings gifts, a tradition in this part of Germany. She was dressed in a white gown, like an angel, with a wreath on her head that was lit with candles. The angel came in and wished everyone good health and happiness and blessings. For a child, it was the most magical experience to see her.

We exchanged gifts and shared our different holiday traditions with each other.

One of the German holiday traditions was for children to put a pair of shoes outside the front door. Anyone in the house who had done bad things would get coal and switches in their shoes when St. Nicholas came by, and the ones who had been good would get candy. To make Lori's night even better, Jackie got the coal in her shoes. There Jackie was, crying, and we tried to console her, "Oh, it's okay! Somebody's gotta get it!"

Jackie: I got switches! You shoulda got 'em, Lori! She was the naughty one. I was the good girl. I shoulda got the candy.

I agree with the girls though. It was the most memorable Christmas of all.

President John F. Kennedy

Not all days were like Christmas days. Like on November 22, 1963. I was off duty at the time & it was in the evening hours, when a neighbor from up above us came running down the stairs to tell us that President John F. Kennedy had been shot and killed back in the states.

He was rattling off comments that different people were making and was really just distraught over having to give us the news. Lori had been in the stairwell herself when he passed her by in a rush. We got everyone together and went for the TV.

Now, when something of this magnitude happens, the field of intelligence, and the military in general, goes on heightened alert. We didn't know to what lengths or how deep the issue was, so first things first were to simply go on high alert, just until they could figure out where this incident was headed and what the implications would be.

The significant thing about this was that not only was the incident covered on German television but there was also an unbelievable outpouring of grief in Germany over the United States & this terrible tragedy.

Lori: It was remarkable to see the general public in a foreign country react so strongly. In most respects, you felt removed when you were away from the states but no…not this event. And not here.

Most intelligence workplaces where I worked during my tours of duty had no windows. I guess it's like submarine duty. If you work in this environment, you are usually drawn to windows. When I was off duty or at home, I'd spend a lot of time just looking out of a window, preferably a big picture window. The alternative was to go outside and just gaze around. I imagine the neighbors thought I was a 'snooping busy-body!'

Well, one day, a German boy who had known the previous residents of our Fischstein apartment noticed me looking out the window. He came to our door and explained his friendship with the previous residents, which led to us getting to know his family shortly after. Come to find out, his father had been severely wounded and disfigured while in combat in Russia on the frontlines. Although blinded, he made it back to Germany by God's grace. Although, he was so disfigured that he never worked in public following his return and hardly left his home. His wife worked and they got compensation for injuries and disabilities.

They invited us over to their house & I just remember meeting his father. I told him that he was forgiven, at least by me. That regardless of the wrong of his country's leaders, he was loyal to his country's calling. He cried, I cried, and I still have strong feelings when I think about it.

Take a moment and just think of this man's sacrifice. He did what he thought was right for his country. This was a soldier-to-soldier emotional kind of moment. I count that as a notable & noteworthy life experience to meet somebody like that, regardless of the circumstances.

We took advantage of our time in Germany as a family. Just like Japan, we embraced the different cultures by taking trips to other nearby towns & countries whenever we could. Although, we had really tight travel restrictions & protocols to adhere to. Travel protocol meant that all of our names went on a roster and that roster would be checked as we departed and once again upon return. This ensured we were accounted for at all times.

Sometimes, depending on what was happening in the world, we might get notice not to travel into a certain area or location. And if you got a notice like that, you just didn't go there, period.

When we went to Berlin, we had to go on either an American Flight or the American Duty train. Lori always joked that all the trains we took were bugged – which, in hindsight, they probably were. If & when we traveled to Berlin by air, we could only travel within the corridors of the US, French & British corridors.

Lori: Mom was the catalyst for our family vacations in Europe. She'd frugally save up funds & then by way of camping, we were able to see most of the Western European countries. The summer of '65 was my favorite. We packed our station wagon & headed for Barcelona. We camped on the Spanish Riviera for almost a month. We made friends with 2 Algerian French boys. We watched the locals eat black spiny sea urchins and we showed them how to make palomitas, using the latest invention, Jiffy Pop. Nothing like it…sea and sand & living in a bikini!

Mary: Germany was wonderful. I loved Germany. I loved their food, the people, the beauty of it, the old things that were built years and years ago. And we just made a lot of really great lifelong friends.

I, too, loved Germany. And that wherever we traveled, we were welcomed.

Beach days on the Spanish Riviera,
the kids with new friends.

After three wonderful years with my family, I'd be wrapping up another great assignment with professional military and civilian personnel, & our time in Germany would come-to-a-close. Yet, unbeknownst to me, one of the most dramatic events of my life was soon to occur.

Combat Development Command
Washington, DC

But first, a short one-year assignment in Washington, DC.

Our intelligence portion of the Combat Development Command was located at Arlington Hall Station in Arlington, Virginia, which is actually considered a suburb of Washington, DC.

```
                    MISSION:
   The CDC's mission was to attempt to postulate
 both the military organization and capabilities
  of a foreign power that could possibly be an
   enemy force deployed against US forces in the
   future. We developed friend and foe modules on
  what we considered the advances that could be
   made in X number of years into the future.
```

This data was incorporated into a war game-type exercise, a computer program if you will, in which the postulated outcome would disclose any possible weaknesses in our own capabilities and/or equipment and likewise in the hypothetical enemy forces. This data was then studied by command HQs and various production organizations that were closely allied with the US military complex.

This one year in DC was a *most* interesting assignment and one-of-a-kind opportunity to work in the realm of future challenges and planning. By the end of this tour, I was promoted to Warrant Officer.

It was the early middle of the Vietnam War by now. The buildup had begun in late '66 & '67. Everything was pointing to a continued build up and that we were going to win this war! More divisions started getting deployed. The President & staff had agreed with Westmoreland that they were going to need a couple more divisions. And with that, I *knew* that my time was coming.

I had a friend with me in DC, a Sergeant Major. He also told me that the quota was going to be increased for personnel. So, I contacted assignment personnel myself & asked, "Do you have any idea what my next assignment might be?" And they said, "Well, since you haven't been to Vietnam as a

Warrant Officer, you will be at the top of the list. You're also coming up on assignment so it will most likely be Vietnam."

To which I replied, "I'm ready!"

Without even being asked, my answer was yes.

Yes, for sure.

Mary: But he volunteered! Can you believe that?!

Mary's right. I technically volunteered myself for the Vietnam War.

As soon as I knew that the list was indeed increasing and the number of people they needed continued to grow, I decided it best to tell Mary. That she might as well prepare for me going soon.

Lori: We (the children) were not part of these conversations at all. I remember listening as my dad propped mom up on the kitchen counter and told her the news. And yes, he brought out the charm.

It went something like that, I suppose. Probably my pulse increased a little to tell her. That I'd probably have at least 30 days of quick schooling, a short leave at home before I shipped out for a year.

Mary: I had to take care of the kids. Where was I going to go, I thought. Just overwhelmed. Kind of frightened, too. But I knew I didn't stand a chance to say no. It had to be a yes. I ended up saying 'OK' in the end.

Michael Sullivan Jr., Lori's Son & Arthur's Grandson: Were you frightened, Grandpa?

Not at all. I was never one to be afraid, maybe concerned. I was willing to serve and anxious to serve. And although expected, I was just really excited. Everything was positive at that time. Without hesitation, regardless of any circumstances, I was prepared to say yes.

It went back to my feeling of being 'shortchanged' during the Korean War. Most of my buddies were there and I wasn't. Because of my training & a few other things, I was held back from going to Korea during the war. It was kind of like, if you were going to school to be a doctor, well, you became a doctor. In my estimation, the highlight of being a soldier is being in combat. *Combat is the true test,* I thought. *Could I really do it?* I knew it would be good for my career. And it was always something I wanted to test myself on, that it would be good to see how I performed. *Could I perform under the stresses?* I wanted to prove myself, *to* myself. Although intelligence work is something different in that you are always in combat, I wanted to prove that all my training was not for naught.

I was anxious to be able to go. And with this kind of buildup, I wanted to be there, ya know? And if I was going to make a career out of this, this was how to do it.

I knew I had a family, but I also knew just as well that *I* had a family at home that could shore up their life while I was gone.

About a month or two later, I got the call. It was the year 1967, just about the middle of the Vietnam War when I officially received my next assignment to Vietnam.

First, I received a month's schooling on the situation in Vietnam, at NSA. This included a detailed education on the structure of enemy combat forces in North and South Vietnam. It also included the disposition of all allied forces, i.e., American, Australian, South Korean, and South Vietnamese (ARVN) units. Emphasis was placed on disposition of forces (friend and foe) in central South Vietnam. That was where I would be assigned with the US 25th Infantry Division. It was a great indoctrination course that prepared me well for this next assignment.

US 25th Infantry Division
Vietnam

I was off to Vietnam in the summer of '67. As for my family, Mary and the children left Virginia to live with Mary's folks in Park Rapids.

Lori: They thought they could just send me to school…with a quick goodbye & an 'I'll see you in a year.' I sobbed into a towel. I threw that towel on top of the cupboard and decided that I wouldn't take it down until he got back. (I only shared this with my Grandma Ruth). I just continued to sob at school. I was in 7th grade. There was no compassion. Maybe they thought military brats should be able to handle this sort of thing, or maybe that children are more resilient than adults.

Now, in case you are not familiar with this 20-year Vietnam War, it was a battle of two philosophies, democracy and communism. Democratic Vietnam against Communist Vietnam. It was the Vietnamese people against the Vietnamese people, a civil war.

I was assigned to a special intelligence company in direct support of the US 25th Infantry Division, a division HQ, in their tactical operations against the enemy. The 25th Division established its base near Cu Chi, a provincial capital between Saigon and Tay Ninh, an area named the 'Iron Triangle.' Part of our division was deployed at Tay Ninh and the other at Loc Ninh.

The division consisted of soldiers, tanks, artillery, and support units. I estimate that there were approximately 15,000 troops in the main division compound. We also had what were called fire support bases in the division's area of operations (AO), which was like a battalion of infantrymen that would have a few artillery batteries, maybe a tank, and the rest of the men, support troops, like cooks, medics, & engineers. They could fire artillery into the surrounding areas, if need be, but their job was to secure their given area, surrounding their fire support base.

```
                    MISSION:
   Our unit mission was gathering intelligence in
   direct tactical support to the US 25th Infantry
   Division. In each of the fire support bases, we
     had two or three men doing some of our intel
work, and another 5 or 6 on a certain assignment.
   In total, our intel company was around 200 men.
```

When I got there, I found it to be just as they said at NSA. No surprises, but still, it was a learning experience. Vietnam is like many other nations where the weather is very HOT during the day. It could be wet or it could be dry, but always hot. So, they fought more in the evening & made their attacks or started an assault in the evening hours or early morning. On my second night on the job, we got hit. And we got hit *hard*. Enemy forces fired a number of rockets into our camp.

This wouldn't be the only attack either. On one of the next attacks, a rocket came in and hit one of our bunkers that didn't have a top on it. There were nine guys in there, all from the finance unit of the 25th infantry division. All of them were killed.

For the next four weeks after these first few attacks, anytime anyone was off, the off-duty personnel would spend time filling sandbags and building roofs on their bunkers. For our bunker, we made a roof out of logs and then covered it with sandbags, so that if a rocket hit the sandbags and exploded, hopefully there would be no casualties.

In the Vietnam War, there seldom was a frontline or a rear echelon. It was kind of a fluid dispersal of troops. Some of the biggest battles were fought inside or near metropolitan areas or near a base camp of a division…like ours. And usually, they were launching 106mm rockets into our camp which were Chinese design and manufacture. Wherever it hit, it fragmented into shrapnel which caused considerable casualties and destruction of equipment and property. The enemy employed a rather crude missile aiming procedure. They would dig a hole in the ground, estimate the angle to where a launch would land, and then fire. After a while, they got pretty good at it. Even if they fired into a camp of 10,000 people, they were bound to hit something or someone. Whether a productive hit or not is another thing.

During an attack, the protocol was that wherever we were at the time of attack, we had to get to an assigned bunker with our weapon, our gear, and our ammo. Only then could we get information about the type of attack, such as actual assault or a rocket and artillery attack. The division then might ask us to supply a few men for the outer perimeter but otherwise, we just defended our own unit position.

One rocket that came in hit our motor pool of all things, destroying three of our vehicles completely and the shrapnel flattening just about every vehicle tire. No one was killed that time, but we certainly had transportation problems for a few weeks.

Then, there was the time that a rocket came through and hit right inside *my* hooch, right by the door. The Lieutenant just across from me was hit and so was another guy at the other side of the hooch. As for me, it blew my mosquito netting off my bed, depositing shrapnel into my mattress but somehow never hitting me.

Mary: One of the very first letters Artie wrote from Vietnam said that a shell had gone through his hooch, and that if he had sat up, it would have killed him. It upset me so much and I was so scared. I wrote back. And in the next letter he told me he would never tell me anymore details like that ever again.

And she's right. I never did that again. It was a regular occurrence & frankly, too disturbing for the family. To be honest, the only thing that resulted was worry, and that's not beneficial to anyone.

Me, outside my hooch in Vietnam

I was the Operations Officer at Cu Chi for the first six months. Cu Chi was the heart of the tactical activity during the early part of the Vietnam War. The troops had good morale and we were kickin' butt every time there was a battle. Once the 7th North Vietnam Army (NVA) division replaced the Viet Cong, which were the South Vietnamese Communists, they became even more effective and knew how to coordinate forces & attack a given target with much more success. In a sense, this allowed us to counter much better with our intel collection because their military units were more organized, traditional in structure and tactical procedures.

At first, we tried our best to have day and night shifts. My job started around noon and I would work until midnight or so, reporting to the command center where our main division intelligence was located. The intel operations personnel worked at times that corresponded to the enemies most active times of day, so they could take appropriate action and report on things in a timely manner. Those that were out in the field would also report in, so we would be collecting information that way, too. Then, we would feed that information to our supported commands for their appropriate action.

Regardless, it was always a 24/7 situation. No time clock. Certain members that worked support sections had the more normal shifts, like the cooks, supply, and motor pool, for example.

One month, I was our Pay Officer, which brought me back to one of my favorite pastimes. Flying!

But I didn't have to pilot this time. I was the one being transported, and by way of helicopter. I don't remember how many places I had to stop to pay the troops, but they were always scattered about in different areas, and it was scary at times. Sometimes when I went out during a mission, the enemy'd be firing at us in the chopper from down below the whole time. We'd have our guys with their guns on both doors, ready to silence any incoming fire & defend the aircraft if needed.

I flew to many places during my time in Cu Chi, even when I wasn't Pay Officer. Each time in the air, a different mission, but always business in nature. It was either to consult, talk something over, or take something back, kind of like a courier service but for intel information purposes. The helicopter was the best way to get places and quick, too.

On one flight, I went in to pay a unit, really out in a grubby area. I paid a guy who was sick. I asked him what was wrong, he said he thought he had the flu. By the time I got back to my base, they said he died.

Apparently, the night before I got there, a rocket came through & hit the pole that supported his tent. Everyone was supposed to be in their bunkers, and he was not, which was a violation. Well, a little piece of shrapnel about the size of a pencil tip of lead went through his kidney, unbeknownst to him. He saw a doctor, but he didn't detect anything wrong. The shrapnel, so small, entered without a trace, passed through his kidney, poisoning him, and causing him to bleed to death internally.

In early December 1967, an opportunity for rest & recuperation (R&R), was available, and I asked Mary what she thought.

Mary: And I asked him NOT to come home. It was just too hard to say goodbye again and again. At this point, I would rather keep it sustained and wait until a time when he could be home for longer.

…however, I wanted to take my R&R at home!

Mary: He said that if I didn't let him come home, then he would go somewhere else like Australia, Bangkok, or Hong Kong! Fine, I said. I'll let you come home.

…And so, I did! For about a week or two. I had been five and a half months in Vietnam. I *needed* some R&R. When I left Vietnam, it was 110 degrees. When I got to Park Rapids, it was -20 degrees, dead of winter.

Mary: In a way, when Artie came home, the children resented him. It had been a different family structure for some time now.

Lori: We were used to living with grandparents! And then Dad came home as 'an officer in command.' He talked differently. He was a bit of a stranger.

Yes, Lori. The authoritative Dad *had* to come home and set things straight!

It was wonderful to see Mary again, my kids, and family. The only thing that tempered those feelings was that I knew it was only for a very short time and then I'd have to go back.

Mary: And I would have to let him go. Again.

After Cu Chi and my time at home, I was sent to Bien Hua. They established a new corps support intelligence unit at Bien Hua, near Long Binh, South Vietnam.

MISSION:
Provide tactical support at the corps level, in this case II-III Corps, incorporating new procedures that allowed us to report valuable intelligence faster & with higher accuracy.

That was a new venture for all of us, support at the corps level versus a division level like I had been thus far. It was a new concept of support. We had been incorporating some new procedures and new abilities to report faster, improve our accuracy of information, and we needed this processing center to do that. And Bien Hua would be this processing center.

The Operations Sergeant had been an instructor with me in Fort Devens, a good man, so I was glad he was there. My Officer in Command (OIC), I appreciated, for letting me put into practice what I knew to be best, and he didn't argue with me about it. He depended upon me to know and implement the technical procedures and for him to take the command-and-control part of it as he should, and this was a true blessing to me. 71 men worked for me in our section, most of whom I got to interview, & then based on their highest and best abilities, placed them where I thought they would be the most productive.

Since day-to-day operations were continuous, and I mean 24 hours per day, there was a matter of sleep disturbance that we wanted…*needed*…to mitigate. So, we got a group of engineers to come in and build a large officer barracks building with separate rooms. And then being in a new place, we also weren't quite 'protected' yet which meant any other spare time was spent filling sandbags. Ultimately, when we weren't at work, we all pitched in where needed & we made sure to do the best we could with the time we had.

There were many days I didn't feel like we accomplished a whole lot and then there'd be two or three days where we were able to produce a lot of valuable intel for our supported US, Australian, and ARVN combat forces. It went smoothly for us. And thankfully, I can say that this whole operation to provide tactical intel support at the corps level greatly enhanced our operation during the rest of the Vietnam War.

Things looked okay for my second half of my time in Vietnam. We had good latrines and bathroom facilities, showers where you always had hot water which was something we all learned to treasure. Out in Cu Chi, we didn't have hot water. We had to make it ourselves! They erected towers and put big barrels painted black so the sun would heat them. The first 10 or 12 people would have hot water but it was all cold water thereafter. I'd shower during the off hours. Even a cold shower was a luxury over there.

The mess hall was huge and the food was good. They would truck in a whole bunch of civilians, most of them women, who were brought in as cooks, servers, and table waiters. In a sense, it was just like living downtown.

Every two or three weeks or so, the company commander and some of the officers would come get me and we'd go to the Bien Hua Air Base Officers

Club for a steak which was a real treat. Now, I don't know how they got those steaks, but they were *really* good.

Mary: Meanwhile, a man had come to the house to see some antiques of my mother's. He found out where Artie was stationed and said, "He is never gonna make it home alive." Then, I ran into someone on the street one other day who said the same thing, "He's never gonna make it home." Hearing this, I would just go home and was really depressed at times.

Lori: Mom was often in bed in the middle of the day. As a daughter, it's a sight you don't forget.

The Tet Offensive

Toward the end of my tour in Vietnam & at Bien Hua, I made a *bad* decision.

It was Tet, which, in Vietnamese means New Year. Two of my old friends were back at Cu Chi, an officer I knew and a warrant officer that had been a Sergeant at the instructor school back in Fort Devens. Things had been relatively quiet in II Corps, so I decided to go to Cu Chi to spend New Year's with them. A getaway, you could call it. Some downtime & to see some old friends of mine. I got permissions from all the right people and took off in a chopper.

Gosh, it was so good to see them! Being at my old place and seeing everything from just a few months ago, although it felt like much longer.

And then the North Vietnamese launched their infamous Tet Offensive, which was when the North Vietnamese & Viet Cong struck the South Vietnamese in a series of planned attacks, targeting areas with heavy US troop presence.

I was now trapped in Cu Chi.

And 'downtime' meant that I had a pistol on me and that was it.

There was rocket fire everywhere, booming and crashing all over the place.

My friends found me an AR15 and lots of ammo. I felt a lot better with this protection.

For the next two days, we were in the bunkers. Not knowing if or when they were going to hit us again. We had an artillery unit, 5 inchers I think, right next to us, firing right over us & across our compound at the enemy and *ohhhhh man!* I'm pretty sure that's what caused most of my hearing loss.

We couldn't go anywhere.

Which meant I was unable to get back to my own unit at Bien Hua at a great time of need.

I kept trying to get back and I mean, I tried everything. They even called in from my home base telling me to catch this heli' over at a certain place. We called the heli-pad every day and, "No, there's nothing going that direction and we can't get you on anyway." At that point, I needed priority to get on. And although I was high on the list, there were also evacuees and the transport of troops including those who were wounded. There wasn't room for me or anyone else.

The only ones that were still going back and forth were the convoys, heavily armed convoys. I'd have to wait to catch one of them…that was going in the right direction. Boy, when I was finally on one, I was a happy guy. I'd never been on one before! And this one sure was interesting…

"Where you from?"
"Where you goin'?"
"What do you know?"

There I was, riding in a convoy, where everyone onboard had a weapon & every other vehicle had a machine gunner on top. It seemed like the sky above us was full of gunships, just looking and waiting for something or someone to shoot at. I remember looking out over the side and just thinking, my God, this convoy could kill an entire army! For the 50, maybe 70 miles back to Bien Hua, it was quite the sight. Nobody fired on us on that trip. You would've had to be a fool to even try.

When I got off the convoy, a driver from my unit met me with a Jeep and I went back through where the Tet Offensive had attacked Bien Hua. Enemy bodies hung over the fences, young people, I'd imagine 15 or 16 years old. There they were, for what seemed like acres. I remember some of them with their yellow arm band markings and then those with green. It was the biggest defeat they could have ever had.

"Glad you're back Mr. Burns!" And it wasn't a matter of *don't ever do that again*. It was more like *we won't let you do that again*.

They were doing me a favor, after all, letting me go like that. But we all figured things seemed pretty quiet. We should have taken that as a bad sign, not a good one.

Although I felt wonderful coming back, I also felt guilty. My gosh, that's when I was needed the most. My first question when I got back to my base, "This isn't over, is it?" And they said, "Oh yeah, in America they think we lost this thing!"

In reality, the North Vietnamese didn't score ONE victory in the Tet Offensive, not *one*. The news media in the US overemphasized the results of the North Vietnamese Offensive.

We were really busy in the days after. We started studying everything for lessons that we could learn from the event. We learned a lot of things about the event, by backtracking and looking forward all at the same time, seeing if there was anything we missed.

What *did* we miss? What *can* we learn from this? At the very least, we could officially say that complete silence or any lull in activity should be nothing less than a warning, and from then on, understood as a threat indicator. Every unit at every level went through this analysis as it pertained to them & implemented new procedures for going forward.

Looking back, we noted that there were a lot of funerals going on a month before Tet. Come to find out, they were having these 'funerals' so they could carry all the weapons in the caskets and bury them in their cemeteries! When the time came on the Tet, they went to the cemetery, got all their weapons, and so it began.

Nobody caught this until afterward, with interrogation and investigation. Human intel (HUMINT) should have found out some of this activity but most of us weren't in a position to be watching people out on the streets and in the cities. We were engaged in completely different intel processing.

There's always indicators, but the importance is to recognize them and do something about it.

Mary: In the meantime, Artie didn't write! I was used to getting a letter every single day. But during this time, The Tet Offensive, I went to the mailbox every day, and never a letter. Five, maybe six days went by. Nothing. And here I am, hearing about all the fighting over there. And it sounded like the North Vietnamese were winning!

I wrote to Mary every day throughout my military career. And if ever I knew that I was going to be gone somewhere and unable to write, I'd write a letter ahead of time and give to a friend to put in the mail at the appropriate time while I was away.

Mary. He did do that.

Mary didn't know that I was trapped in Cu Chi during the time of The Tet Offensive. For that trip, I thought, well, I'll be back in a couple days. Little did I know…

Lori: There was so much anxiety in the house at this time. We watched the news, and it made us so worried. I remember Mom not sleeping well. One late night, early morning, I rode my bike into town to get the very first newspaper so we could get the latest news.

Mary: I remember when I finally got a letter. The mailman was used to giving these letters to me. So, after a while with no letter, he actually found one in his bag that he had missed. He came all the way back and brought it to me. I'll never forget that. That was so nice…I could have hugged him.

In a sense, this tour was a test, proving ourselves to our commanders that yes, we can really do this, and to myself, that *I* could do this. That it may be a lot of work and come with a lot of challenges that I've already since forgotten, but I know that they served me well. The highest medal I received in my military career was at the end of this tour.

Around this time, when someone is about to leave and go home, a soldier makes his or her sweat chart, a calendar of some sort that you checked off every day that you made it back safely. Everyone had a sweat chart. Most never started one though until they were thirty days out. Some guys made designs around theirs and some other things of which we won't be going into great detail here.

But there are always tales you hear. *The guy who was supposed to leave tomorrow but got killed today.* Or, *'a soldier was leaving on a plane to go home and got shot down and killed.'* I don't know how many of those tales were true but once a soldier got down to five days on their sweat chart, there was lots of anticipation to say the least. And at about that time, they'd report out of their unit, turn in their combat gear, and move out of their barracks to an out-processing unit.

When it was my time, they sent me from Bien Hua to Long Binh, where I got to spend the night with the commanding officer and sergeant major that were friends of mine. We got to relax and had one drink together. The sergeant major had a bed made for me for the night and the next day they had a chopper arranged for my departure. This was a very nice gesture, respectful, and appreciated very much because I could have just taken a truck or something but traveling by chopper was always easier.

There's also of course the feeling when that chopper takes off. The wheels lift from the ground and you're in the air. You think, in that moment, that no matter the circumstances of what happened then or what was now, you made it. You were going *home*.

The Legion of Merit & The Bronze Star Medal

Coming home is always a husband and wife thing. It has to be that way. Mary left the children with her mother and she met me in Minneapolis. We went to Detroit together and picked up a brand new car, Ford LTD, with the lights that opened up and all that fun stuff. We spent a couple weeks together, just us.

Getting home to the children…was awkward. They still sort of resented me like the first time I came home. And anyway, it's different. Here I was used to getting up myself, taking a shower, going to eat, and getting to work. Whatever my routine was. Then, all of a sudden, I go home and there's my wife and my family.

These few days are now known as a readjustment period and there are now counseling services to help with those kinds of transitions. But not then. Before, it was a two-week ship ride for a soldier coming home, which provides that soldier with a window of time to readjust by his or herself. Now, a soldier is home within hours of being in combat, which, in some cases, is traumatic in itself.

For me, I got up in the morning and didn't have to take the mosquito netting off my bed or put it back on at night. Little things like that were just really strange. You wouldn't think so, but it just is.

NSA
Fort Meade, Maryland

My next assignment was in Fort Meade, an assignment where I got to come home every night, indicative of a stable time period of our lives. School was close-by as well as hospitals and facilities. We were on base. Things were *okay*.

At first, it was just Mary & I, getting settled in. When the school year finished, we went back to Park Rapids and collected the children. Our new neighbors in Fort Meade at one point said to us, "We thought you were the quietest people we'd ever met in our life!" Think again, neighbors.

The kids would quickly miss Park Rapids and the feeling of 'home.' By then, Grandma & Grandpa had spent a lot of time loving on them and had very little discipline with them, too. Now we were a family, 24/7. With time, we got through it.

It was a good tour at Fort Meade. I was assigned to NSA as a Section Chief, supervising 16-20 people involved in intelligence activities. About half were military and half civilian analysts. I worked on the very same problems pertaining to the Vietnam War that I had in Vietnam and, as far as work challenges go, it was a smooth transition to which I adapted and enjoyed to the fullest.

We were in Fort Meade for three years. Things with the children started to settle again and got better. We established some normalcy the very best we could.

And then, I volunteered to go back to Vietnam.

In this occasion, it was interesting. I knew I was coming up on my 3 years' tour at Fort Meade and I was already at NSA. So, I just called the assignment section and said, ya know, *I'm coming up again and I want you to know that I would like to go back to Vietnam, if possible,* because that's what I've been working on. That was the target for me now for the past two assignments. It was also the shortest overseas tour available at the time and then I'd be back home, I thought. And as it seemed, my request was in & it was going to work out that way.

Korea was the other option, which was 13 months without my family *and* no combat pay, working a target I had never worked & knew nothing about – bad deal. It just would have been difficult. I much preferred Vietnam.

Wouldn't you know, in two weeks I got that assignment to Korea.

The big boss of NSA came by one day, "I hear you're leaving us?"

I said, "Yeah, it's pretty bad though, I'm going to Korea."

"Well, how would you like to go back to Vietnam? Represent us in our HQ in Saigon?" he said.

"Well, that can't be, they've already cut the orders."

"If you want this assignment, I'll change the orders."

"I can't think of anything I'd rather do than that, sir."

It wasn't a week later when my assignment to Korea was revoked by the order of the Department of Defense. I called assignment but the Warrant Officer who tried to send me to Korea wouldn't talk to me. In fact, he never talked to me again. But the boss of NSA? He could change any orders he wanted.

I was awarded the Joint Service Commendation Medal at the end of the assignment in Fort Meade.

Joint Service Commendation Medal, May 1971

I was officially going back to Vietnam for a second tour. And it would be an assignment like no other, before, or after.

Next order of business?

Tell my family.

Mary: He volunteered! He begged me to let him go back to Vietnam. He kept after me so much that I finally said OK.

The begging of Mary the second time...man, I really had to do some talking. Let's just say I was Mr. Nice Guy for quite a while.

Mary: We moved back to Park Rapids. Artie & I bought an apartment house. There were four apartments, three were rented out, and we moved into the first-floor apartment. I was going to live independently this time.

Lori : I remember this move to be the hardest. I was a junior in high school. I was close in age to many fellow students who were being drafted. It was an extremely emotional time. There was just too much to process. I had to say goodbye to my dad again, to leave a wonderful school in Maryland, and I had post high school education to think about. Being in a small town had its pluses and minuses, the biggest plus were the grandparents.

Lori had a hard time, watching 'Dad' go the second time. It was the adjustments that were hardest for her. Different schools, switching back and forth. Jackie & Tommy did okay, though.

Lori: We went from going to school with like-minded kids -- whether it was a school on an army base or military base or Maryland, where it was filled with NSA kids and military kids, going through the same thing, more transient, the pick-up-and-go type, all able to make friends easily -- home to Park Rapids, a little small town. It was UGH. More than going overseas & more than going to another country, moving back to a small town was the biggest challenge. Luckily, I found a friend who also moved to Park Rapids that spring, Ione Bomstad.

This time, I left them in the bitter cold of a Park Rapids winter.

UNDERCOVER
Department of Defense
Saigon, Vietnam

Saigon, Vietnam, February, 1971.

I received documentation as a civilian GS-12 working for the Department of Defense (DOD), Special Representative to Vietnam, 'DOD Spec Rep' for short.

I kept my real name, but the rest was fake.

I'm told it's called 'shallow cover,' or undercover. It was a brand-new thing for me & a brand new thing for them, too. It was the first time at this HQ that they took a military man and civilianized him into a civilian government worker position. I was the first one chosen for the job, in support of an entire military complex in Vietnam.

In a sense, it was an honor for me to be selected.

Being an undercover spy…in Vietnam? Well, I felt positive about that. I thought it could even build up my career bio so to speak. All a plus as far as I could see. It was a thrill, really, to believe that it could all turn out beneficial to all.

I showed up in my uniform but had to stay in hiding until they could get me 'civilian' clothes. *So…just hang around in Vietnam?* I thought. *What am I to do? Hang around where? Hide where?* All I could do was sit in my hotel & wait.

All was done after about a week and I was introduced to my Vietnamese counterpart, Major Tah. I was presented to Major Tah as 'Mr. Burns' from NSA, a civilian GS-12, his equal in rank in his army.

The Vietnamese have this thing. If you were a colonel and I was a captain, it would be losing face for me to tell you what to do. In reality, I was a warrant officer, which was 'below' his rank in his army & something he was never allowed to know. So, for my working relationship with Major Tah, I had to be presented as equal rank. If Major Tah ever got promoted, well I got promoted too, even if it was only fakery. If he ever did know, then he hid it well.

Nonetheless, Major Tah was a brilliant man and very interesting. He had previously worked with the French, so he spoke beautiful French, English, Chinese, all the dialects of Vietnamese, and more. We worked together & bonded well. We talked first about what our assignment was and what our relationship was to be. I asked if I could have a desk in his office, to which he quickly assured me I could. I learned to be careful and follow the Vietnamese routines instead of our own so as not to be disruptive in any way.

MISSION:
First, to pass technical information and intelligence to Major Tah from various US sources that were cleared for Major Tah to receive. I would be Major Tah's only source on this level and this kind of information. Second, assist Major Tah in establishing enhanced delivery systems of intelligence to their supported commands.

Say we had a target and a US command wanted to know about it. I would go and ask him specifically, *"What have you guys got on this unit or target?"* But that target would have a different name. The name he had would be different than the name I used, giving the intel back to Major Tah. I always had to be careful with that.

We had to establish communications that were secure, efficient, & timely. Their ethics and procedures and the way they did things, were slow and cumbersome. But there was a war going on. Anything that could help the war effort, we needed to be able to get it to the supporting commanders and fast, not by carrier pigeon or some guy taking it up on a motorcycle. There were better ways.

Where another nation's army can be in comparison to ours, was shocking. I didn't know how they did as well as they did. We needed to improve their communications and security systems which may sound like an easy job, but the things we ran into...

In most foreign armies, communication has to go up a line, a hierarchy if you will. I spent a lot of my time meeting with in-between people in order to meet with the people in authority that I actually needed to meet with. They'd do so much lip tisking and eye rolling and hmm-ing, start talking in Vietnamese with their counterparts until they decided, finally, after all their mannerisms, that we could talk to the next level. The end goal for us was to get to the general who would sign off on or nix the problem at hand. Something that should take a week took a month at best.

Major Tah was very knowledgeable about other cultures, so this was nothing new to him, other than watching its effect on me. Likewise, what I was introducing to him was new for him, too.

Major Tah had probably 100 men under him. They did everything by pencil, no computers, and no air conditioning. We tried our best to get them to modernize their procedures. Some of the communication equipment, such as the encryption devices, had to be temperature controlled. We had to turn on the air conditioning in whatever room the equipment was. But as soon as you put a Vietnamese in an air-conditioned building, they would put on coats because they were cold.

After we would leave that location, within a week, they would often turn off the air and sure enough, they would burn out the equipment! After burning through not one but several pieces of equipment, we had to come up with an educational-type program for them, to ensure that if they want to make headway without this crazy constant battle, then they would have to follow these procedures we put in place. It was a cultural learning curve for both parties.

We battled those things just as much as we celebrated progress. One moment I was discouraged, wondering how long was this going to go on? And the next day, an ounce of progress. We saw their intel functioning well & their information to the supported command, timely. When we saw this, this told me, yes, we can do this. It might be challenging but with determination, we can do it. And indeed, we did it well. With time, Major Tah's group was very capable and productive.

I traveled with Major Tah when we visited the Vietnamese Army bases but alone when visiting the US intel sites. On shorter trips, we went by helicopter. Farther away places, we flew with a contract air service called *Air America,* which was funded by another US intel agency. They had phenomenal aircraft that could get you in and out of the most difficult landing spaces I've ever experienced.

I wrote Mary a letter every single day while I was in Saigon and if I knew I was going to be gone for days at a time, I wrote 3 or 4 letters at one time & dated them with their correct future dates so that she wouldn't know I was gone. Because very quickly, I started flying *everywhere* in Vietnam, to every place that the ARVN were engaged in our type of intelligence work, from Vietnam's Demilitarized Zone (DMZ) in the North, to the southernmost city of Ca Mau.

Me & one of the helicopter pilots

A Distinguished Guest of Honor

The Vietnamese always wanted to treat us, wherever we went. Whether a dignitary, a ranking person, or a member from our headquarters, they wanted to celebrate and honor us, as the Americans who came over to help them. To help establish their government and maintain their freedoms. It was an opportunity for them to celebrate and commend the work we had done.

They'd host these banquets and special ceremonies for us as their 'distinguished guests of honor.' At every one of these small banquets, with Major Tah & me as the distinguished guests of honor, we were treated with the very best food and the very best drinks.

Later come to find that the guest of honor would receive the skull of whatever meat we were eating. It could be the head of a chicken, head of duck, or it could be the head of a rat. The skull would be somewhere beneath a big bowl of gravy, big chunks, and served in a big dish for each person at the table to serve from. Whenever Major Tah escorted me to these events or just went with me, he'd take his chopsticks, serve me my portion of whichever head-of-animal we were having and then serve himself second.

Thankfully, Major Tah was also there to protect me. I had informed him in advance that I was *not* going to eat these honors. So, he'd slip the skull of the whatever off my plate & somehow dispose of it.

Another time, I had to be careful about offending the others at one of these banquets. I picked up a piece of the meat that was on my plate and started chewing it. Then it got hard. Which was just when I realized that I was chewing on a skull.

When no one was looking, I took it out of my mouth and looked down at it, only to see its eye sockets staring back at me. I nudged Major Tah and said, "Please let this be squirrel." "No, that's rat!..." he responded excitedly… "…very common!" So, I just held this half-chewed rat skull in my hand and wrapped it up discretely into a napkin. I then gave it to Major Tah and he smiled. Culturally, I had to get through that.

At another banquet we were invited to, they were eating these little pieces of meat, white on the outside and red on the inside. I had never seen that before, so I didn't take any. Most of the US guys around me were eating this stuff, ooh-ing & ahh-ing over how delicious it was. So finally, I couldn't help but to ask Major Tah what it really was. Major Tah told me, "Oh, that's hog bowels. They eat the guts of the animals, too! They wash them, cook them, and then eat them!" One of the US guys that ate this stuff got an actual parasitic infection of the stomach, a parasite *so* bad that they had to send him back to the states for medical treatment.

As for me, whenever I had to eat any kind of animal over there, I always washed it down with 33 Beer, pronounced 'Bam-E-Ba' in Vietnamese, which was a common Vietnamese beer that contained formaldehyde. Can you believe that? I thought, at least it would kill whatever it was that I was eating. I'm sure it wasn't too good for my digestive system, but it had to kill or embalm whatever was in those animal parts... *and* probably saved me!

Mary: The last thing I told him before he got on that plane to Vietnam, don't eat any risky food and don't fly. He did both.

But I had to!

Mary: I know you did, Artie.

Really, though, to be a distinguished guest of the Vietnamese army, was truly an experience in itself to appreciate.

Typical banquet to honor visiting guests,
Major Tah is on my right

146

Saigon was the capital of Free Vietnam, or South Vietnam. We built a huge embassy there, with army and marine units close by. We even had a row of houses for ambassadors from other countries near the capitol. They had their capitol, their senate, and all the government buildings, with a huge complement of security to protect that area of Saigon. It was like having our very own Pentagon. We had a lot of civilian technicians, like engineers, electricians helping the Vietnamese to improve their factories, their businesses, their power plants, infrastructure, transportation, their everything. This included all the US military service including the South Korean and Australian forces, even the US CIA.

All that to say that it was hard *not* to see an American on the streets during the day. Simultaneously, we'd see the Vietnamese in their military uniforms…riding around on their bikes. I often wondered…what were their duties really? If we were there to support them in bringing back a viable honest government, then their troops should have been in troop compounds where they were supposed to be, being paid as active soldiers and air force mechanics, not riding around all day on bicycles.

I think every alley had some kind of little manufacturing shop, run by families and their children. They'd be hammering on metal all day, turning our metal scraps from battle into sellable items that would later become collectibles.

I thought, *so this is war?* Other than rockets they fired into the city for harassment, that was it. Very little was happening in the city. And in my opinion, that wasn't a good thing.

Suffice it to say, my second tour of Vietnam was proving to be a really interesting education, just as predicted.

The Time I Was Most Frightened
My Sunday

To be honest, very few times in my life, regardless of my career, have I been frightened. I don't know why. There are certainly times where I should have been frightened but I'll recount this one very specific frightening experience.

During the week, everyone would be out doing all kinds of different things. Sometimes we'd be together, around headquarters dealing with our counterparts in the Saigon area when things were calm and peaceful. Other times, we were in other parts of Vietnam doing classified things. But Sundays, if we were in Saigon, those were our days to unwind.

We'd make our reports for the week and contact who needed to be contacted & then, it was downtime. Some of us who worked together would go to this beautiful old French hotel in the middle of Saigon. They had a splendid café at the hotel where they made coffee and French bread and baked goods. It was a treat. I mean, come on, freshly made bread with butter and coffee on a Sunday morning? There was nothing better.

And we always went in the morning because there would hardly be anyone on the streets at that time. We'd load into a car, at least four of us, sometimes in a Jeep. And whoever was driving was responsible for that vehicle. The driver would drop the other three off at the cafe, then go park the vehicle in a military protected site, put a chain through the driver's wheel and secure the vehicle. It was about a block and a half from the lot to the cafe. The driver had to go around the back of all these buildings and then make their way to the front of the hotel.

On this particular Sunday, it was my turn to drive and my turn to secure the vehicle a block and a half from the hotel. We had the Jeep that day. I let the guys off at the front of the hotel, there were some other cars unloading too. Then I went out back to park the Jeep, chained it up, and started walking to the hotel. I was walking fast, you know, out of safety first and the desire for the fresh bread, butter, & hot coffee that awaited me.

As I was walking, I turned the corner of a building and walked into a leper. Face-to-face, only within inches of each other. He was dressed in a black cloak with a hood draped over what was left of him. The sides of his nose exposed. His cheeks, mouth, hands only degrees from rotting away.

I jumped back & yelled in pure shock.

His hand was out, palm up, & he looked me right in the eye. The flesh of his hand was wet, the tissue just barely covering the bones with some fingers already partially gone.

He begged me for money, probably for food or further transport away from the leper colony that he obviously had escaped from. And if I recall correctly, I darted around him.

At first, I was mad. I was mad that he'd ruined my day. My *Sunday.*

Did I touch him? I kept thinking. *Did I get too close?*

Next, I felt ignorant. I could have given him what I had in my pocket and maybe that would have lasted him a month on the street or at least until someone caught him and took him back to his colony.

And then, *why couldn't I have just backed up?* Instead of yelling and running away like I did?

And last, I felt ashamed. All my fears had rapidly turned to anger, anger at myself.

I made my way to the hotel & to the guys back at our table. I told them what happened & then they were all concerned looking around. We had our coffee and our rolls, but in a situation like that, your taste buds go awry. *My Sunday.* It was not a pleasant day in the slightest.

I didn't understand leprosy much. I knew it was contagious but not at the level that I had been so fearful of. I started thinking about how Jesus touched a leper, because He had a greater compassion, a greater understanding of sickness and death. That was probably my first real spiritual awakening, in hindsight.

I was still mad at myself though, for a little while, wondering what I should have done or could have done. I responded to this incident out of pure ignorance and regretted my reaction very much. I was left in pure turmoil. And for the first time that I recall, I was dreadfully afraid.

After some time had passed, I concluded that maybe the Lord let this happen to me to get me thinking about life. That we can be here today and gone tomorrow.

Am I compassionate enough, I asked myself? Or, *do I put my own skin first?*

Why did this happen to me?

Why did I become so fearful of a victim with leprosy, when rockets had come into my base, only inches from taking my life, not just once but multiple times, and I wasn't fearful?

In the end, the most frightening experience of my life had nothing to do with war.

My Conclusive Thoughts on The Vietnam War

When I left Vietnam, we were right at the critical point of the South Vietnamese winning or losing the war, and I honestly thought they could do it. They still needed a lot of American support, in order to become a strong, viable, and successful country that could also maintain its democratic identity. But it was going to cost a lot of money to get them there. The United States was still supporting their army and their infrastructure, their police forces, things like that. There were many other agencies that were involved, that even I don't know anything about to this day. From what I could see, I thought they were going to make it and I believed in my heart that they could.

To my chagrin, five years later, the US pulled the plug, and abruptly left Vietnam without a respectful, peaceful agreement. They simply abandoned the South Vietnamese without any agreements for civil treatment of the populace by the North Vietnamese.

Disappointment at its highest.

I felt that it was too much effort and too much work gone down the tubes, right before my eyes. To this day I haven't been able to visit the Vietnam Memorial in Washington, DC. It's just too hard to come to terms with. It felt like a personal tragedy had occurred. Many, many lives were lost, and you have to say to yourself, for what? We should have never started unless we had the commitment to complete this to victory.

Let it be known that we never lost a battle in Vietnam. Not one battle. The newspapers came out in those final years, *we were losing and getting our butt kicked every time we turned around so let's get outta there, it's not worth it.* That wasn't the case at all.

We were so close to a complete and total victory, democracy coming back, good leaders getting put in place, good generals getting in charge and taking charge of their armies. They should have stayed until an honorable peace was achieved at the very minimum. The government of South Vietnam was to be the recognized government of that country and the war would end, period. Anything less than that, unacceptable.

I'm not advocating anything other than the idea of being committed from top to bottom. Isn't that what Dad said you oughta do when you're plowin' the field? You don't 'half' plow the field. You plow the whole thing!

The ending of the Vietnam War went against my very nature. My beliefs, everything.

I love America.

I hope that we recover the status we once had. I hope that we can learn from our history. That's what history is for, for us to benefit, learn, and grow from.

I hope we recover the hope we once had.

I hope that our motive will be to stabilize the difficulties of the world without military action.

I hope, above all, that our leaders prioritize being truthful and honest with the people of our nation. What you say, as the leader of our nation, well, you better honor that.

Call me an oddball, I don't know, but it seems so logical to me. The fear of the Lord, law, order, and discipline make for good government. And that true peace, justice, and prosperity will follow.

If a man or woman says *I'm gonna go and serve my country for peace and freedom of the world,* he or she is a hero. It's a different commitment in the heart and in the mind. There's this thought process that unless you died defending your fellow man, then you're not a hero. No. All men and women that agree and strive honestly to be good soldiers, even onto death, are heroes. Of course, we highly honor those that have fallen - but it does not undermine the work, the roles, or the commitment of our other fellow service men and women.

Major Tah & his army issued this award to me for my achievement and
technical advice to them.

Pictured is a personal present that was gifted to me from Major Tah,
"Mr. Burns, this is to you from me.
My personal thank you for all your support the past year."
These are hand-crafted Vietnamese murals of fine art,
his 'thank you' for all I did with him.

> **The last time I saw Major Tah was in Saigon right before I left. They had a big celebration for those that were leaving, another banquet. Five years or so later, when Saigon fell, I got a call from a Colonel I'd worked with. He told me that Major Tah, who was then a Colonel, was assassinated by the North Vietnamese.**

Overall, I enjoyed my second assignment in Vietnam very much. The challenges, dealing with different people, I liked that part about it. I went away feeling at least I'd done every possible thing I could do to make things work.

My work as DOD SPEC REP included many trips in & out of most of the ARVN military posts, big & small. I was awarded my second Bronze Star medal for this assignment.

NSA was pleased with my work, having been a trial period of a military person occupying a civilian position. My replacement arrived and was also an Army Warrant Officer. However, they did civilianize him prior to his arrival, so at least he didn't have to hide before his cover was established like I did.

It was a little easier coming home from Vietnam the second time around. Since I had done it only three years ago, I kind of knew how the emotions went and what to expect.

Up until the time I left, the assignment officers kept telling me that I was going to Okinawa, Japan next. I really didn't want to do that.

I was coming up on 23 years in the military and the children were starting higher education, so I thought of getting out and starting civilian life.

I didn't fight for it though. I just figured I'd ignore all the talk of Okinawa and see how things played out.

Well, just before I was going home from Vietnam, and thinking Okinawa was next, I got an alert that stated I was going to Heidelberg, Germany. *Hmm, well, I'd really like to go there!* I couldn't really think of what we had to offer over there or what kind of set up was in place. And that's because they were, in fact, setting up something new.

New seemed to follow me just about everywhere I went.

US Army Headquarters Europe
Command Support Group
Heidelberg, Germany

And, the family could come with me! Mary was so excited. We made such great friends the first time around so the thought of meeting up with some of them again was something to look forward to. We both loved the culture, the community, and on top of that, the schools were ones we could count on. All in all, Mary & I were elated for a second assignment to Germany.

The one drawback, it was right in the middle of the school year for the kids. Mary and I decided I would first go alone to Germany and she would follow with them after I had confirmed housing on base. In the meantime, they'd be at our apartment house in Park Rapids.

As usual, Mary began with her protocols. She sold our car, gathered the passports, school records, immunizations and all the other requirements for her, Jackie, and Tom.

All the while, I went to NSA in Fort Meade and got briefed on the Cold War situation in Europe, especially the military situation in the Eastern bloc.

Once I secured housing, the family flew to Fort Dix, New Jersey, and then on to Frankfurt, Germany. Lori stayed in Park Rapids. She graduated from high school, stayed the summer with Grandma and Grandpa Johnson, & then was off to college that fall. While going through that time in her life, it was really hard to leave Lori in Park Rapids.

It was a new special intelligence unit called Command Support Group (CSG), reporting directly to the commander of Allied Forces Europe. The Soviet Forces Germany, the Russian Forces that were still stationed in East Germany, they were the recognized threat of the day in free Europe. The challenge at hand was quite simple and singular.

MISSION:
To establish a viable intel center in direct
support of the US Army Headquarters Europe. Our
unit, CSG, dealt only with special intelligence.

There was a full Army Intelligence (G2) Center there. This was the top US Army Headquarters for our forces assigned to the European Theater of

Operations. This headquarters was commanded by a four-star General and full of brilliant people.

They had a desk just on infantry, one just on tanks, one designated desk for just about anything. And everyone manning those desks was at least a Major. Most were Lieutenant Colonels and the head of the whole G2 intel section at this HQ was a full Colonel. We had a full Colonel as a Special Security Officer (SSO) to the General for our special intelligence material.

They had plenty of intelligence resources covering the enemy, so you'd think they wouldn't need us, right? Well, the general felt that there was a sliver of special intelligence that he wasn't getting and that he wanted. Furthermore, the Army HQ also felt that the receiving wasn't timely enough. So that's what we were there to develop, or the gap we were there to fill.

Our CSG consisted of a chief (Lt. Colonel), an NSA rep, 4 Warrant Officers, and approximately 25 enlisted men. It was a new assignment, a new mission experiment, and I was glad to play a part in it. I was excited for what lie ahead, despite any challenges that may come along with it.

We had the ability to communicate directly with most of the high-level intelligence centers in the US and Europe. Because of that, we needed technicians to work around the clock, making sure that the caliber of intel support would not only be the best, but completely operational 24/7.

By this time the challenges of establishing a new intel unit were no longer 'new' to me. These challenges followed me around and I liked that about my career. At the end of each day, I just hoped that we could pull off whatever mission was at hand. And here in Germany, that it be a respected, honest to goodness intelligence supporting unit, the best that it could be.

First, we went through the throws of development. We published a full intelligence report every…single…day. It was a tremendous amount of work. Our special intel units located in Europe reported items of interest directly to us and were added into the reports to our supported command.

For me, this adjustment was the exciting part, the fact that we could get so much information and so quickly. To be able to find an answer to a question or provide an answer to the supported command requests in real time. Definitely new to me, an enhanced capability that we now possessed.

But amidst our progress, there was a concurrent change. The situation in Europe was changing rapidly. The Russian influence diminished steadily. We knew something was up. The Russians were still there and their intent, still obvious. But the elements were changing, politically and militarily, in both our government and the Russian government. There was a lessening of tensions and the opening of honest dialogues between leaders and for once, it wasn't just for show. West German leaders, England, US, Russia, the intent felt sincere! It felt like, *let's lessen the threat of the military and just pull back, let's at least try to get along, let's at least try to make for a more peaceful world and here's how we can do it.*

As steadily as the Cold War situation changed around us, we also remembered our past experiences and the lessons learned. That in lulls or quiet times, we must remain prepared. Things can change rapidly, never *ever* let your guard down. That we must continue to provide our commanders with the most accurate up-to-date information possible, on both the enemy's intentions and their capabilities. That was forever our effort.

Generally speaking, I felt that progress was made in our work efforts, in most areas, in what we were there to do. The world looked like it was going to be safer and more cooperative, more supportive of one another with a better attitude, better relationships.

I traveled a lot while I was in Germany. Militarily, I went to Rota, Spain to attend a conference at a naval base, close to Gibraltar. To Harrogate, England where I was fortunate to pass through London on the Queen's birthday, pageantry at its highest. And to many of our intel sights in Germany, including Berlin.

We also traveled a lot as a family while in Germany. But our travel decisions were often based upon the gas rations we received from the military. We only got so many gas stamps that we could use outside of our own base camp's gas. These stamps were honored in other countries by the US Status of Forces Agreement with these countries.

We traveled to France, Italy, & Austria. We always went over for the Flower Festival in the Netherlands, acres of tulips amongst what seemed like millions of other flowers. They'd make garlands and put them on the front of your car.

My assignment in Heidelberg was great for many reasons, but one being that we could incorporate these kinds of special trips into our yearly, sometimes monthly living. The children were always excited to go wherever we decided. Not to mention, Germany sees many gloomy, dreary days so we enjoyed venturing south for those that were sunnier.

During our assignment in Heidelberg, we had two of the most pleasant situations. Like the fact that our visits resumed with our German friends, The Kunz family! We exchanged many visits with them during those three years on the second tour.

We also saw my fellow soldier comrade, Bill Gilbo, who was Army active-duty service from our National Guard unit in Hibbing. At the time we were stationed in Heidelberg, he was stationed with his family in Nuremburg, about 80 miles from us. He was the HQ Command Sergeant Major of the main military unit of that area. Many visits were exchanged with them, too. Interestingly, they were in Heidelberg during our first tour in Frankfurt. And now here we were stationed in Heidelberg while they, in Nuremburg.

During the last year of my assignment, I started thinking about re-assignment. What and where it could be, but also what I'd want to do. I reflected on family circumstances, especially those of my children and the next steps of their education, advanced education by that time. How more movement of the family would and could play into that equation.

Simultaneously, there was a big push on all Warrant Officers to get their college degree. I had two years of college credit by then, but I would need to complete college to get put back up for re-assignment and promotion. This meant potential separations from my family if I went that route. All in all, that option just wasn't looking good.

Mary loved travel and the different locations that came along with our military lifestyle. Despite everything, she was perfectly willing for us to stay in if I wanted, which made that part of the decision totally mine. For me, though, I could look down the road and see the different factors in play. It ultimately came down to the fact that going any further would be too complex, and the impact on our family life, too demanding.

One of the things in most of the wars of the past is that most of the newly drafted were young. The military buildup of these wars consisted mostly of young single men and women. In most cases, these newly inducted personnel engaged directly in war activities and were discharged at the end of hostilities. This situation gradually changed after WWII as more individual soldiers began choosing the military as a career. The short peace periods between WWII, Korea, and Vietnam certainly were a factor in this cultural change.

After WWII, there wasn't much in the way of military family support. Only after the Vietnam War did they realize that the hardest thing on married people was the concern over family stability. Many didn't have places with support for their spouses and children, like we did. Some families really had to go through difficult family times to make it to their retirement.

In the case of Mary & I, one of our kids was already in college and the next two would hopefully follow soon. This, coupled with the need for a 2-year college assignment, followed most likely by another overseas assignment, convinced me that *now* was the time to retire from the military.

THE RETIREMENT ASSIGNMENT

M ary agreed. That after our time in Heidelberg, *it was time for all of us to retire from military life.* In July of 1974, I submitted my retirement request.

The Colonel in charge of our unit in Heidelberg asked me to reconsider. Although I did for a moment, I still turned down his request in the end. He put me in for The Distinguished Service medal, which I received just before my departure from Germany. He was a fine Commander – one of the best.

I was honored and pleased at my discharge, words aforementioned, and The Meritorious Service award presented to me by the Colonel at my discharge.

Jackie & Tommy flew home alone from Heidelberg in September of 1974, off to Grandma & Grandpa's so they could enroll just in time for the new school year. They were 16 & 15 years old respectively so that also meant they were authorized to travel alone! Not a concern being as they both knew how to speak up if there was trouble & how to fend for themselves. On their orders it stated, 'unaccompanied dependents of CWO Arthur Burns,' which meant they followed a special set of procedures from departure to arrival. Lori and her boyfriend, Michael Sullivan, met them at the Minneapolis Airport and drove them to Park Rapids.

In late November, we shipped our car home & soon after, received confirmation of its arrival in New York, ready for pick up. But before we headed back to the states, we spent one last very special Christmas in Germany with Kathy Kunz & the family.

On the first of the year, we flew to New York. We picked up our car, visited some friends in Virginia, & headed for Fort Sheridan, Illinois, about 28 miles north of Chicago, my final military destination for discharge & retirement processing.

I still had a month and a half of active duty left. When we arrived, they said, "Well, we don't know what to do with a Warrant Officer…" Because you don't just send a Warrant Officer 3 out to sweep the decks!

"So…you go home and you call us on the first of every month and tell us you're alive. We'll send you your discharge. Unless you want a parade in your honor?"

"I'll skip the parade," I replied.

"Alright. You will be placed on the retired roll on February 15, 1975."

Mary & I accepted and considered this just a new assignment in the USA.

Homeward bound we drove until, due to an ice & huge snowstorm, we became marooned in Menomonie, Wisconsin. As a matter of fact, we were stuck only 14 miles from the place I was born. It really was a big snowstorm, ice on the roads, the whole nine yards. After a few days, enough roads and ramps had been plowed and we were able to make it home, just in time for Super Bowl IV. And again, the Vikings lost.

Anyone got a spy job around here?

We still owned the big apartment house in Park Rapids, so we notified our manager & asked him to leave the bottom apartment vacant. We moved in once our household goods arrived from storage. After settling in a little more, I planned to start looking for a job.

Anyone got a spy job around here? I thought. Can you believe there was nothing available in the Park Rapids area!

My aunt, Florence Hensel, who passed on Sept 13, 2023 at 103+ years, ran the employment office in this area. She asked me one day, "Well Artie, are you looking for a job?" Not knowing what else to really say, I said, "Sure, I guess…!"

Only 4 or 5 days after I officially retired & came home, Aunt Florence got me a job in a lumber yard.

Something new & something I knew nothing about. Another big challenge.

It was similar in that it was like a new station, or a new assignment, with a challenge that I wanted to win, overcome, and also enjoy, just like I had everything else. If I made a mistake, then I'd do better at my next go at it. There really was so much similarity in moments like this and throughout my life, that reminded me of the disciplines of military life that serve a soldier well all their life. You have to stay structured and focused. It doesn't matter the endeavor of your choosing, so long as it gives you purpose and meaning in life.

I learned about lumber, deliveries, trucking, and all the terminology that comes along with that. Working at the lumberyard also gave me the opportunity to observe the trade of construction. Most of the building trade in our area consisted of 1 to 4 men teams that dealt with wood construction and then would sub-contract to the electric, water/sewer, heating/AC, and landscaping trades for complete building of homes and other types of buildings. This served me well. Because in June of 1976, the 200-year centennial of our nation, Mary & I bought a lot along the Fishhook River in Park Rapids.

For Mary & I, this started another work of love. Up to this point, we'd been in & out of one military home after another, across the globe & back again. So after a full, transient life in the military, this home in Park Rapids would be our first single family home that we would ever own.

It was a vacant lot, so we got to build! Together, we designed our 3-bedroom, 2-bath split-level house that would overlook the Fishhook River. There was just something about being on the water that we desired.

We had a construction team erect the shell of the house, including roof, windows, and siding. And just like I had learned, we subcontracted the electrical, plumbing, and heating. We then hired my cousin, Matt Hensel, to complete the interior finishing and trim work. We also hired a stone mason to install a stone fireplace in the game room of the lower level.

This lot was located on the very edge of town, so we also had to put in our own water well and septic system, which we did ourselves with help from family and friends. Another friend helped Mary and I to design and construct the rooms and single bath on the lower level. Soon after that was done, Mary & I moved into our 'home sweet home.'

The next year we built and completed our attached 2-stall garage and about half of the exterior decking. The next year, 1978, we built the rest of the decks and covered porch. In 1980, we built another garage, unattached, & with 3 stalls. The following year we completed the landscaping, which was followed by dock, lakeshore development, and design the year after that.

Building our home was a challenging yet fulfilling and creative endeavor. As of now, the rest is just periodic updates and repairs that can be surprising at times!

Our Home

Around 1982, I quit the lumber yard. And I actually started going to school for carpentry. Realistically, I wasn't going to start a carpentry business this late in life. It was just for self-improvement and to learn some of the basic skills of the construction trade. But I did also find myself searching for new employment, too. And my next opportunity, well, let's just say they found me & I found them.

Downtown at the courthouse in Park Rapids, they were looking for a Veterans Service Officer (VSO). In the state of Minnesota, the state mandates that each county must have a VSO to serve the veterans that reside in their respective counties. The role really caught my interest, so I applied for the position. Fortunately, I was selected and held that position for the next 11 years.

As a VSO, I assisted veterans who needed to file for federal compensation for service-connected disabilities, ensured they received the proper medical treatment at VA hospital locations, & secured any required transportation in support of these actions. There were also a lot of state benefit packages available for veterans and their widows in some cases. There were state and federal retirement and nursing facilities that were and still are available to qualified veterans.

It was a complex job that required a lot of scheduling tasks for the first two years. But for the 11 years I was VSO, it was very rewarding to be in a position to help the veteran population of my home area of Park Rapids and Hubbard County.

Veterans Service office meets clients' needs

By Sue Blom

The Veterans Service Office is set up to perform a variety of tasks. They include:

*Effective administration and supervision of all duties and functions of the Veterans Service Office, in accordance with all county and appropriate state and federal procedures.

*Initiate and prepare claims, and assist veterans in presenting claims for benefits from the Veterans Administration and the Department of Veterans Affairs.

*Assist veterans and dependents in obtaining evidence in support of a claim such as affidavits from former service buddies, or other persons to substantiate a claim for servcie connected disability, non-service connected pension or other veterans benefits.

*Prepare claims for increase in service connected compensation because of increased disability or increase in number of dependents.

*Assist eligible veterans to gain admission to a Veterans Administration Hospital.

*Work closely with the lcoal Sunset Nursing Home to secure full benefits for veterans and/or their dependents that are residents at this facility.

*Work closely with the local clinic and hospital relative to VA assistance in medical payments. Also in the transfer of veterans to VA facilities and/or nursing homes depending on condition and projected treatment of the veteran.

*Assist eligible veterans with transportation to Veterans Administration Hospitals.

*File applications for eligibile veterans for outpatient medical treatment and outpatient dental treatment.

*Advise and assist veterans on all phases of government life insurance.

*File applications for eligible veterans for education under the GI education bill, vocational rehabilitation and veterans educational assistance program.

*Contact and assist survivors of a deceased veteran in obtaining death benefits such as burial in a national cemetery, burial allowance for funeral expenses and plot, government grave markers, government life insurance proceeds and survivor pension benefits.

*Assist a veteran or eligible survivor in obtaining VA home loan benefits.

*Initiate claims against service departments or other agencies for claims such as travel allowance for veterans and dependents, claims for property lost or damaged during shipment by a service department and errors in pay, discharges and personnel records.

169

PARK RAPIDS ENTERPRISE SAT., NOV. 8, 1986 - PAGE 2A

Burns' career began in 1949

By Sue Blom

For Art Burns of Park Rapids, Veterans Day is a time to recall past military experiences, and explain his current support role.

A native of Park Rapids and graduate of the Park Rapids High School, Burns found his military career beginning in May, 1949, when he joined the local National Guard outfit.

"As it turned out, I joined the U.S. army in December, 1949. It was just a short time after World War II, and it was the thing you did. Back then there was a lot of pride in our nation and in serving the nation.

"I had graduated from high school, and the economy was not too good at the time, so I signed up with a friend. I got 'talked into' the intelligence area because it sounded interesting. After all, doesn't everyone want to be a 'super sleuth'?"

Burns attended language and classified schools, and although he is a Korean War veteran, he admits he was in school for most of the time until about the close of the war. However, upon completion of school, he spent three years in Japan.

His career in the army led him to see two tours of duty in Vietnam. "That was probably the ultimate assignment. The war was an unpopular war, and because I was working with classified information, I was privy to so much the general public never knew. This made

things frustrating, because I couldn't even tell my family what was going on."

Burns was part of the 25th Infantry for his first tour of duty in Vietnam. "I was right in the combat area for at least seven months. It proved to be rewarding in a manner of speaking. Being so close to the action, I was able to see the intelligence used to save lives, right away. It's not often you see the results of your work in the intelligence business pay off before your eyes within minutes."

Burns' second tour over to Vietnam was with the Department of Defense. "I worked under a civilian cover, and ended up visiting almost every base involved in the war. It was at this job that I worked with the 'important people'. By important people, these folks word changed entire policy involving the war."

"Looking back, he revealed, "I had two very different tours of duty, supporting the same war. I think this is true of a number of veterans from the war. At different times during the period, different attitudes were prevalent.

"The Vietnam War was the first media war. A number of events were highly distorted. For example, the TET Offensive of 1968, which was a major victory for us, was treated by the media as a major defeat.

"In addition, there was a growing anti-war effort coming to light in the United States. Until 1967, public opi-

nion was pretty good about the war, but then the chemistry of the times took over and things didn't work out. I, personally, feel the U.S. was justified in entering the Vietnam War as the Korean War and World War II.

"The feeling about the war had changed so much from my first tour to the end of my second tour. The last time I returned from Vietnam we were advised not to wear our uniforms because of the strong anti-military sentiment in the country.

"It's unfortunate because it was the individual soldier who had to bear the brunt of the opposition. When the government who sent you to war starts criticizing the action itself, the soldier wonders why they are even there."

Following his duties in Vietnam, Burns received a new target assignment in Europe. "I was involved with setting up new intelligence support for the 7th Army, part of the NATO forces. We designed our own communications channels.

"I also spent three years stateside, serving as an intelligence instructor. This was rewarding in the sense you saw someone come in with no knowledge, and leave ready to take part in the nation's intelligence operations."

Burns looks at his years of service and the variety of

Burns

Cont. on page 3A

Art Burns is a familiar face around Park Rapids. Following his retirement from an Army career, he has been the Hubbard County Veterans' Service Officer for the past four years.

PARK RAPIDS ENTERPRISE, SAT., NOV. 8, 1986 - PAGE 3A

Burns

Cont. from page 2A

duties he was assigned. "They were all so different. The time I spent in Japan was during a period of Soviet missile exploration and development.

"Family-wise, the time spent in Europe was great. We were together and had opportunities for the children to become fluent in foreign languages. We also had the opportunity for travel there.

"As far as strategical intelligence, being in combat and able to do a job, deemed well done by my superiors, is very gratifying, especially when you see your work save lives."

Burns retired from the Army in February, 1975, ending a career of military service, and starting a life in the civilian world. His honors include Retired Army Warrant Officer, Legion of Merit, Meritorious Service Medal, three Bronze Star Medals, Joint Services Commendation Medal, three Army Commendation Medals, four Good Conduct Medals, Vietnam Cross of Gallantry with palms, Vietnam Service Medal, Armed Forces Reserves Medal, Meritorius Unit Citation and National Defense Service Medal.

"When you retire at the age of 45, you're too young to sit around and do nothing. I some different things, and then about four years ago the job of veterans' service officer became available."

All of his experiences have helped Burns in his role as the veterans' service officer for Hubbard County.

"I was there, and I know the problems soldiers faced. Although I am a Korean War vet, I'm close in age to the World War II vets, and I saw the actions in Vietnam, so I can relate to most of the veterans we have today."

He noted, "The veterans from the Vietnam era are the most difficult because fighting took place over such a long period of time, and the mood changed from the beginning to end. A number of soldiers, when they returned, did not want anything to do with the whole situation because the nation's public had turned their backs on them.

"Also, because of transportation, it was very common for a soldier to go from the civilian life to the front lines within 24 - 48 hours. This was also true of the discharge. There were no staging areas to prepare soldiers for the front, or to prepare them for entry back into the civilian world.

"If anyone should feel guilty about the Vietnam War, it should be those individuals who criticized, or didn't go. It should not be the ones who served their country, or gave their lives."

As service officer, Burns is responsible for making sure the need of veterans are met. These needs range from information, education to pension payments. There are approximately 2,500 veteran, widows, orphans and retirees in the area. Presently Burns is working on establishing a veterans' council.

"There is a renewed pride in America. I think a lot had to do with the fact there was no hero after the Vietnam War. But there has been a rebirth of patriotism."

He added, I enjoy working with men and women from the service, and I want to see them get all they are due. We're getting good support from the county board and the local service organizations. This can only lead to improved service to the area's vets."

Fourth of July: *Parade's grand marshals are three retired veterans*

from Page 1A

Firecracker Foot Race

The 36th annual 5K Firecracker Foot Race starts at 9 a.m. Friday, July 4.

It's best to register online at www.firecrackerfootrace.com. Online registration closes at 8 p.m. today (Wednesday), July 2. As of Tuesday morning, 532 people had registered. Organizers are estimating more than 700 people will participate.

Participants are encouraged to provide ample time to get ready for the race. Bib pickup is the night before from 6:30-8:30 p.m. Thursday, July 3 at Heartland Park and at 6:30 a.m.

Pickle Events will be timing. Refreshments will be available following the 5K.

Residents in the East River Drive neighborhood should plan accordingly as racers will make their way through the neighborhood between 9 a.m. and 10:30 a.m.

If possible, limit vehicle traffic in the area during this time.

The event is a fundraiser for the Park Rapids Area High School boys and girls cross-country and track teams.

Pie auction

Beagle Books & Bindery will host the 8th annual John Michael Lemna Pie Baking Contest and Auction this Fourth of July.

There is a $5 entry fee per pie. The categories are Single Crust, Double Crust, Custard/Cream and Juvenile (for bakers ages 8-12). Cash prizes given to top pies in each category, including best in show.

All pies minus one slice will be auctioned at 3:30 p.m. in front of Pioneer Park on Main. Proceeds benefit the Park Rapids Area Library.

For more information contact Jennifer at Beagle Books at 237-2665.

Fourth of July parade

The parade's theme this year is "A Hometown Celebration."

The parade will march at 5 p.m. with the line-up beginning at 3 p.m. on Monico Lane. The procession will head south on Main to 5th Street (Beatty Used), turning west to Heritage Living Center, where it will disperse.

The parade begins with a military line-up, the Ameri-

PFC Lloyd A. Stigman

can Legion Post 212 Color Guard, Marine Corps League, American Legion Auxiliary, Sons of the American Legion and Disabled American Veterans with Boy Scouts assisting.

This year's parade grand marshals are PFC Lloyd A. Stigman, CPL Algene (Dude) Bomstad and Chief Warrant Officer 3 Arthur Burns.

PFC Lloyd A. Stigman entered the U.S. Army Feb. 14, 1944 and served in the European Theater of Operations.

He was awarded the Purple Heart, European- African- Middle Eastern

CPL Dude Bomstad

Campaign ribbon with one Bronze Star, Combat Infantryman Badge, Expert rifleman, Ardennes (France) campaign ribbon and the Good Conduct Medal.

CPL Dude Bomstad entered the Army Air Corps December 1942 and served until January 1946. He served as supply sergeant in a B-26 bomber unit stationed near London, England. He was awarded the Eastern European theater ribbon, American theater ribbon, WWII Victory medal and the Good Conduct medal.

Chief Warrant Officer 3 Arthur Burns entered the

CFO3 Arthur Burns

U.S. Army May 6, 1949 and retired Feb. 15, 1975. CWO3 Burns served two tours of duty in Vietnam in 1967 and 68 and 1971 and 72. Awards and decorations include the Legion of Merit, two Bronze Star medals, Meritorious Service medal, Joint Service commendation medal, three Army Commendation medals, Army Good Conduct medal with 4 OLC, National Defense Service medal, Vietnam Service medal, Vietnam Cross of Gallantry with palm device, Meritorious Unit commendation with 1 OLC, Sharpshooter and Vietnam Campaign medal.

Pre-fireworks fun

Celebrate your patriotism and love of music all at once as the Park Rapids Area Community Band performs a pre-fireworks concert, starting at 7:30 p.m. Friday, July 4 in Heartland Park. The band is moving from Red Bridge Park due to construction.

You won't want to miss a note from the piccolo to the big bass drum. Dr. Russell Pesola conducts the band in a concert of patriotic music and marches that sets the stage for the Independence Day celebration's grand finale.

Grand finale

The day comes to a close with a fireworks display overlooking Fish Hook River and Heartland Park. Beginning at about 9 p.m.

Because the pyrotechnics are set off in Heartland Park, some parts of the park will be closed for safety reasons. There is still plenty of seating, plus space in Red Bridge Park or in a boat on the Fish Hook River.

The show, sponsored by the Park Rapids Rotary Club, will begin shortly after dusk.

After 11 years, I retired from the 8-5-grind part of my life and entered into the *supposed* 'ease-of-life' era...

HA! I've never been busier!

Those of you that have reached this benchmark know what I mean.

To the rest of you, the mystery will be self-evident later.

Burns' career of service will end in February

By Lu Ann Hurd-Lof

For 10 years now, local veterans have relied on Art Burns for information, advice and services.

Burns has been Hubbard County's Veterans Service Officer since September 1982 and plans to retire - again - in February.

He retired earlier from a distinguished 26-year military career.

When he was a boy entering high school, he spent hours plotting the American Army through Africa and Europe and in the Pacific Theater of World War II. "It set the stage for my life," he says. "I always wanted to be a soldier since I was a kid and it worked out that was exactly what I got to do."

If someone plotted Burns' career on a map, it would look like a post World War II history lesson.

Burns and several of his classmates joined the Army in 1949 after graduating from Park Rapids

High School. He went through basic training at Fort Riley, Kansas, and was selected to go to Military Intelligence School at Carlisle, PA.

When the Korean War started, he was put into language school to learn Russian and spent the duration of the Korean War going to language and other intelligence schools.

After the war, he served four years as administrative manager and Battalion Sergeant Major with the Minnesota National Guard at Hibbing.

In 1953 he went back into intelligence work in the Army and was stationed at Hokkaido, the north island of Japan, for three years.

Following that he zigged to gain more instruction at an intelligence school at Fort Devens, Massachusetts, then zagged to Germany for three years, and wound up at intelligence headquarters in Washington, DC.

From there he was sent to his first tour of duty in Vietnam (1967-68). He returned to Washington, DC and worked three years there with the National Security Agency.

On his second tour to Vietnam he was a representative for the Department of Defense, under civilian cover. He did special intelligence work for the Vietnamese helping them improve their own intelligence network.

The final years of his career in the Army, he was at special intelligence headquarters in Heidelberg, Germany. He retired in 1975.

Through it all, not only was his wife, Mary, very supportive and understanding, but their two children had some unique educational experiences, among them learning to speak basic Japanese, French and German.

Burns listed the high point of

Burns
Cont. on Page 2A

Art Burns was decorated with at least 21 medals of honor since he entered the U.S. Army in 1949. (Liz Shaw photo)

11-11-92

174

News

Fourth of July: *Parade's grand marshals are three retired veterans*

from Page 1A

Firecracker Foot Race

The 36th annual 5K Firecracker Foot Race starts at 9 a.m. Friday, July 4.

It's best to register online at www.firecrackerfootrace.com. Online registration closes at 8 p.m. today (Wednesday), July 2. As of Tuesday morning, 532 people had registered. Organizers are estimating more than 700 people will participate.

Participants are encouraged to provide ample time to get ready for the race. Bib pickup is the night before from 6:30-8:30 p.m. Thursday, July 3 at Heartland Park and at 6:30 a.m.

Pickle Events will be timing. Refreshments will be available following the 5K.

Residents in the East River Drive neighborhood should plan accordingly as racers will make their way through the neighborhood between 9 a.m. and 10:30 a.m.

If possible, limit vehicle traffic in the area during this time.

The event is a fundraiser for the Park Rapids Area High School boys and girls cross-country and track teams.

Pie auction

Beagle Books & Bindery will host the 8th annual John Michael Lernia Pie Baking Contest and Auction this Fourth of July.

There is a $5 entry fee per pie. The categories are Single Crust, Double Crust, Custard/Cream and Juvenile (for bakers ages 8-12). Cash prizes given to top pies in each category, including best in show.

All pies minus one slice will be auctioned at 3:30 p.m. in front of Pioneer Park on Main. Proceeds benefit the Park Rapids Area Library.

For more information contact Jennifer at Beagle Books at 237-2665.

Fourth of July parade

The parade's theme this year is "A Hometown Celebration."

The parade will march at 5 p.m. with the line-up beginning at 3 p.m. on Monico Lane. The procession will head south on Main to 5th Street (Bearly Used), turning west to Heritage Living Center, where it will disperse.

The parade begins with a military line-up, the American Legion Post 212 Color Guard, Marine Corps League, American Legion Auxiliary, Sons of the American Legion and Disabled American Veterans with Boy Scouts assisting.

This year's parade grand marshals are PFC Lloyd A. Stigman, CPL Algernon (Dude) Bomstad and Chief Warrant Officer 3 Arthur Burns.

PFC Lloyd A. Stigman entered the U.S. Army Feb. 14, 1944 and served in the European Theater of Operations.

He was awarded the Purple Heart, European-African- Middle Eastern

PFC Lloyd A. Stigman

CPL Dude Bomstad

CFO3 Arthur Burns

Campaign ribbon with one Bronze Star, Combat Infantryman Badge, Expert rifleman, Ardennes (France) campaign ribbon and the Good Conduct Medal.

CPL Dude Bomstad entered the Army Air Corps December 1942 and served until January 1946. He served as supply sergeant in a B-26 bomber unit stationed near London, England. He was awarded the Eastern European theater ribbon, American theater ribbon, WWII Victory medal and the Good Conduct medal.

Chief Warrant Officer 3 Arthur Burns entered the

U.S. Army May 6, 1949 and retired Feb. 15, 1975. CWO3 Burns served two tours of duty in Vietnam in 1967 and 68 and 1971 and 72. Awards and decorations include the Legion of Merit, two Bronze Star medals, Meritorious Service medal, Joint Service commendation medal, three Army Commendation medals, Army Good Conduct medal with 4 OLC, National Defense Service medal, Vietnam Service medal, Vietnam Cross of Gallantry with palm device, Meritorious Unit commendation with 1 OLC, Sharpshooter and Vietnam Campaign medal.

Pre-fireworks fun

Celebrate your patriotism and love of music all at once as the Park Rapids Area Community Band performs a pre-fireworks concert, starting at 7:30 p.m. Friday, July 4 in Heartland Park. The band is moving from Red Bridge Park due to construction.

You won't want to miss a note from the piccolo to the big bass drum. Dr. Russell Pesola conducts the band in a concert of patriotic music and marches that sets the stage for the Independence Day celebration's grand finale.

Grand finale

The day comes to a close with a fireworks display overlooking Fish Hook River and Heartland Park. Beginning at about 9 p.m.

Because the pyrotechnics are set off in Heartland Park, some parts of the park will be closed for safety reasons. There is still plenty of seating, plus space in Red Bridge Park or in a boat on the Fish Hook River.

The show, sponsored by the Park Rapids Rotary Club, will begin shortly after dusk.

175

Since March 1993, our lives have been filled with spiritual growth, fishing in Minnesota and Canada, and traveling the world with family and friends. I would like to try and put as many of these events into some type of chronologized order as I can, with a few of the highlights therein.

Spiritual Journey

To be brief, both Mary and I have always been of the Christian faith. Denominationally, Mary was Methodist and I, Lutheran. We were married in the Calvary Lutheran Church of Park Rapids, Minnesota on the 7th of March 1953.

In the winter of 1979, both Mary and I became born again Christians in accordance with the Gospel of John 3:3. Since that time, life has been easier for us. We have enjoyed and appreciated the good things of life and can handle the hardships a little easier.

My parents and family were some of the first to join and establish Calvary Lutheran during its formation. We met in the basement of the old Civic Center, later the Ringer Building, probably in 1942. I was also in the first class of confirmands at Calvary. Since we didn't have a church building at that time, we held our confirmation ceremony in the Lutheran Stone Church in Dorset, Minnesota.

Something about my Grandma and Grandpa Burns is, I believe that they started the day right. They read scripture and then prayed for specific things in their lives, for their church, area, and the nation. That might be a ritualistic thing, but it sure starts the day correctly. To this day, Mary & I do the same.

After retiring from the service, Mary and I felt that our faith was dry or that something was missing. About this same time a new non-denominational Pentecostal church was forming in Park Rapids, New Life Christian Fellowship. It was led by Fred & Beverly Brown. We attended a few times and were convinced that this was our church. We dropped anchor and are still members today.

The connections made in this church led us on many trips across the US to attend seminars and meet people leading wonderful lives and associated with various projects to help mankind.

Guadalupe, Mexico

Around 1980, a retired couple, The Lunds', began spending part of their winters in Southern Texas. They were volunteer workers with Habitat for Humanity. The following year, they met Pastor Steve Brewer, who presented his plan of constructing a large Christian-based orphanage in Mexico. Miraculously, 27 acres of land were donated to Pastor Steve to build this orphanage. The land was completely barren desert and located approximately 5 miles from the city of Guadalupe, about 15 miles south from the Texas-Mexico border crossing in Fabens, Texas.

Our friends, Arly & Laura Lund, convinced Mary & me to follow them & try it out. For us, we thought it would break up the Minnesota winter while at the same time, we could accomplish something of value in our lives and help the very needy of Mexico. In the spring of 1983, we were off to Mexico for a new type of work vacation.

It wasn't until much later in the winter of '91 or '92, we started to clear the desert area & erect the very first building of the orphanage!

The NW property marker, the building of the wall
to surround the 27-acre orphanage

Another gentleman & me, working on the first building of the
orphanage, the first aid station

During the process of developing & building the orphanage in Guadalupe, many other missionary endeavors were accomplished. The first was the acquisition of a Tapestries of Life (TOL) headquarters site. It started at a temporary building & storage site in Clint, Texas, where we worked for the first two years and lived in a pop-up camper during our visits.

Next, Steve bought a run-down dilapidated ranch site with acreage just outside of Fabens. We repaired the house & converted an old helicopter port into the headquarters office building. We constructed campsites for the workers, dug a new well, upgraded and enlarged the water and sewer system, & rehabbed two other buildings on the lot into livable quarters for the many permanent volunteer workers at TOL. After about 7 years, this became the official home base and TOL headquarters, still in operation to this day.

Later, in Fabens, a large building complex that was a former church was purchased, to be repaired and become the primary housing site for the work groups that came down to work on the orphanage. Volunteer worker groups typically visit for one to two-week periods at a time and have to be self-supporting. This building now includes a large kitchen and dining area, an auditorium, men & women sleeping quarters, a supply storage area, & also an attached manager/family apartment. It took about 5 years to finish this truly great and much needed support facility.

During all this time, TOL simultaneously helped the indigenous pastors of Guadalupe and surrounding areas extending all the way to Juarez, Mexico, the sister city to El Paso, Texas, and the slum areas of Anapra. Every work group that came through TOL usually would set aside one day to help and visit the needy in Mexico to give them a better understanding of the major spectrum of need in the area.

We can talk poverty in America but that does not resemble poverty in places like this in Mexico or other emerging countries. In one place we went, the mother had taken in kids from the neighborhood that were abandoned and barely surviving, working her tail off to help support them all. Some families might have as many as 10 kids living under one roof. And it really does break your heart.

The staff at Tapestries planned to accept orphans and be fully operational sometime in 2023. By this time, it will have been a 30-year labor of love. Mary and I have been richly blessed to be part of this great missionary effort.

Chihuahua, Mexico

Mary and I were also involved in another orphanage in Chihuahua, Mexico. It was led by Refuge Ministries Inc and was more traditional in its development. It started in a home with 4 or 5 orphans but rapidly grew from there. We helped with various US work crews that bought new property & expanded with buildings. It is now a large compound with a large orphanage & separate manager and caretaker quarters. Another enjoyable and fulfilling experience.

Charlottesville, Virginia

In the spring of 1983, Mary and I visited the newly founded missionary headquarters called Advancing Native Missions (ANM) in Virginia. This new mission group was led by two men, Carl Gordon and Bo Barredo.

Bo had been to our church in Park Rapids the previous year. We were very impressed with his sincerity and vision which was to form a support group that would provide financial and spiritual support to native Christian pastors in other countries of the world.

Carl and his wife Minda graciously hosted us in their Virginia home for our entire visit. They included trips for us to the old Maxwell hunting estate, Maxwell House, now a ministry center. The other trip was to the University of Virginia, located in Charlottesville, that Thomas Jefferson established. We were able to bless them in Charlottesville all the way from our church in Park Rapids.

To say this was a grass roots operation would be an understatement. As I remember, they had three rooms on the second floor of an office building. One of the rooms was large and filled with a table piled high with office supplies. A few chairs by the table and one typewriter. As a new Christian I thought, *they really have faith, guts, and a positive attitude,* but they needed a ton of prayer.

Since that time, they have far exceeded their goals and are now supporting and assisting struggling Christian workers throughout the world. What a faith builder this was for Mary and me.

Bo and Marlo, Oliver Asher, Lucille, and many others from ANM have since visited our church in Park Rapids. We have been richly blessed.

From left to right:
Lucille Le Beau, Cris Paurillo, me, Mary, Carl Gordon, Bo Barredo

Our missionary adventures have been a long and arduous challenge but also fulfilling and rewarding in the spiritual sense.

Fishing Life

The next major past-time in our retirement years was devoted to fishing. Here in Park Rapids, we live on the Fishhook River which is connected to four lakes in our area. Minnesota is known as "The Land of 10,000 Lakes." Our county of Hubbard has about 1,000 of them. In other words, there is an abundance of fishing lakes in our immediate area.

A strange oddity, if you live on a lake, you fish the opposite shoreline. And those living on the opposing shoreline fish on your shoreline. Maybe it's a grass is always greener kind of thing…

In this regard, Mary and I decided early in our fishing experiences that Minnesota just wasn't good enough, so our target area turned to Canada. Initially, for about four years after retirement, I went up to Canada with "the guys." In about 1980, we, guys, decided to take our wives along for some real fishing. That was the end of "the guys" fishing trips.

We usually went twice a year, spring & fall seasons. The trips were always for a week, sometimes two weeks. We had our own boats and could go to any lake or river that we wanted to. We went out in the boonies and stayed in tents. As we got older, we decided to upgrade to "resort-style" camping.

We fished the lakes from Kenora to Ear Falls and the English River chain that is connected to hundreds of lakes, including the huge reservoir lake called the Lac Seul, and extends about 150 miles from Ear Falls, south and east to Sioux Lookout. All the lakes and rivers in this vast area empty into either the Hudson Bay or Lake Superior.

These fishing trips lasted about 40 years, 1976-2018, and were honestly more fun having Mary with me. COVID-19 ended the trips to the present. Friends and families that accompanied us on many of these trips included Mary's parents, the Engsts (3 generations), Jim & Janet Walter, two Johnson families, and many others.

The uniqueness of fishing many of the lakes in the remote areas was not only great fishing, but many times we would not see another person or boat the entire day. Just some bear, moose, and other wildlife.

Great fishing, experiences, friendship, and the beauty of God's creation are memories beyond description.

The one that didn't get away

A great day, a great catch,
with The Walters

Ron and Ellie in the boat ahead of us
Beckington Lake, Canada

Zippel Bay at Lake of the Woods, Minnesota
with our daughter, Jackie, & her husband, Jeff

The Shack

"Hunting shacks" are one of the novelties of this area. This can range from a big tent to a mansion. In between lies tarpaper shacks, motor homes, pick-up campers, and all the versions thereof. The optimum is for one to have or be part of a "Hunting Shack."

The first two years after retiring from the Army, our cousin, Steve Johnson invited my son, Tommy, and me to hunt with him at his "shack." And well, we had great luck! My son got a nice doe and I got a buck each of the two years.

It was the shack life that caught my fancy. It added a new dimension to deer hunting, and I caught the fever. Thanks, Steve, for getting me hooked.

I was fortunate enough to find 80 acres of land up in the gulch area, about 20 miles north of Park Rapids, and purchase it with 3 of my cousins, Jeff Green, Lyle Vold, and Steve Johnson, who actually sold his slot to Solon Green, Jeff's father, soon after.

It was an old DNR log building, probably of the 1930s vintage. It was in dilapidated condition, but fortunately Jeff and Lyle were carpenters, so away we went. We built a stone fireplace, a sleeping loft, new roofing, porch, and other rehab. We also, with the help of a friend from the county land department, blasted a large hole in a low spot on our land and created a pond with permanent fresh water for wildlife in the area. To us, it was a perfect castle, alias, "The Shack."

Each owner could invite a friend to hunt during the season. Usually, 6 to 8 people was the average size of our group.

The first thing needed in a "Shack Crew" is a chef to prepare all the meals while the camp is open. Jeff Green was "the man" so that need was met. Another need was firewood for the winter season as no electricity or gas was available in this remote area. We declared a weekend in the early fall months to cut & split enough wood for the winter season. We also did any needed roof repairs and erected deer stands.

A deer stand could vary from a small platform in a tree to a cabin on stilts with roof, heaters, windows, and comfort extraordinaire. The latter is preferable as the hunting season can get sub-zero, and with the wind factor? Very cold to say the least. The object of course is to sit in your stand all day, quietly, and to be lucky enough to get your "Trophy Buck," as early in the season as possible – called "bragging rights."

The stories told, mostly true with some additions, were *legendary*. I wish I had documented some of them. We didn't have cell phones or GPS in those days. Everybody had some family member who got "lost," but not really, just someone who couldn't find their way back to the shack or car. Stories about "The Big One" that everyone had previously bagged under extreme conditions, and then the rack size that sometimes grew with each retelling.

We played card games & even gambled on occasions, but mostly for fun and relaxation.

Tommy never really enjoyed hunting & neither did anyone else in my family, so after about 20 years, I sold my share of the property to another member of the Green family. Now it is a Green Family "Hunting & Recreation Shack." I pray it will be enjoyed and appreciated by their families as long as they wish. I will always cherish the great experiences at "The Shack."

My son, Tommy, & his first buck!

Blessings & Mishaps

A wonderful thing occurred in 2016 when Mary's cousin, Blaise Johnson, who was also our wedding flower girl, remodeled our kitchen & dining room, with design assistance from Kay Stewart, another cousin of Mary's. It was highly needed and our thank-yous do not fully describe our appreciation – we were truly blessed – THANK YOU, Blaise and Kay.

***

The first mishap begins with a not so good story of The Shack. Every winter, a few of us would take in supplies for a couple of days or a long weekend and snowmobile the many miles of groomed trails in the area. We'd party, play cards, and just plain enjoy.

I remember it was about January because Mary and I had taken our taxes and paperwork with us. One day, we each took a snowmobile out to go riding. Mary on one and I on another. There was a big pitstop along the route, where we could get food, drinks, and groceries. We headed there and I went ahead of Mary, who never drove too fast.

We were going down this road together and I sped up ahead of Mary, going over a few hills and having fun. I thought she was behind me, but when I looked back, she wasn't there. I turned my snowmobile around to go check on her.

And I saw her.

She was coming down the same road but on the wrong side.

Which meant, we were both driving on the same side of the road…driving directly toward one another.

She was already coming up the hill and I was about to go up and over it, but at about 65 miles per hour. I tried to hit the bank to avoid crashing head into her but instead, the brush along the road propelled me *and* the snowmobile right into her.

She didn't fall off her snowmobile but she just kind of froze. Meanwhile, I hit her snowmobile, flipped, and up I went with my skis, one of which hit her right in the head, skidding right past her shoulder. Thank God she had her helmet on.

Luckily, somebody drove by, picked her up and took her to the hospital. I had flipped over and fallen on my head, with a cut now, just a little banged up. Somehow, I convinced them that I was okay. Off Mary went to the hospital and I went back to the cabin.

Later, I started walking a little goofy. When family members came, they took me to the hospital, where I should have gone to begin with. Mary was all stitched up & now I was kinda shaken.

That was the end of snowmobiling for Mary. I continued for a few more years, but now, not anymore.

Mary & me,
a snowmobile trip to The Shack, 1979

Onto the second mishap.

It was the last day of duck hunting in the late fall of 2008 when my friend, Ron Engst, and I tipped our canoe while recovering our duck decoys. Ron was able to hold on to the canoe, but I sunk to the bottom. Fortunately, my foot hit something solid and in one final kick, I was able to reach the surface and grab the canoe. We were in luck when a neighbor saw our plight from shore and was able to get a rope out to us just in time. An ambulance arrived and thank God we were both treated for our hypothermia and survived.

Long Lake 'heroes' rescue duck hunters

BY COLE SHORT
THE ENTERPRISE

Two Park Rapids men have been called heroes for their roles in the rescue of a pair of hunters whose canoe capsized Saturday on Long Lake.

Art Burns and Ron Engst were duck hunting on the lake Saturday morning when their canoe capsized, sending them into the icy waters below.

"Someone told us we were leaning over the canoe to grab decoys at the same time," said Engst, 62, a Park Rapids Insurance agent. "I guess the wind caught us and sent us both into the lake."

Jake Beireis was standing outside his home along Long Lake about 8:30 a.m. Saturday when he saw the canoe tip.

As the men yelled for help, Beireis dashed into his basement to grab a pair of chest waders and a piece of rope.

He asked his wife, Shirley, to call 911 while he returned to the lake and furiously flung the rope to Engst and Burns, clinging to the canoe.

"I waded out as far as I could and threw the rope, but I came up short," Beireis said.

In retrospect, Beireis said he should have tied a weight to the rope to help its flight. "In that situation, you're not thinking clearly," he said. He estimated the men were floating in about 10 to 12 feet of water on Long Lake.

Engst said he and Burns tried to swim to shore, but strong winds, along with the chill of the water and heavy clothes, made it difficult. "We tried to kick our feet and paddle as best we could," he said, "but we weren't making much progress."

That's when a second hero arrived, authorities said.

The Rev. Roger Olson, also a resident of Long Lake east of Park Rapids, heard the commotion and came to help.

Olson quickly put a paddle boat he owned into the lake, retrieved the end of Beireis' rope and delivered it to the capsized duck hunters. Beireis then pulled both men to shore.

"They saved their lives," said Tim Archambault, Hubbard County chief deputy, referring to the rescuers.

"They're heroes. I don't think either one of the hunters would have lived without the help" of Beireis and Olson, Archambault said.

See RESCUE, Page A3.

Rescue: *Lake's chill not felt immediately*

from Page A1

Beireis downplayed his role in the rescue. "I did what you had to do," he said.

Engst, meanwhile, called Saturday's rescue a miracle.

"It wasn't just a coincidence that Jake was there and saw the canoe tip over or that Roger hadn't put 'his' boat away for the winter," he said.

"It wasn't a coincidence. It was a miracle. And Jesus is still in the miracle business."

Engst said it was difficult to estimate how long he and Burns were in the icy waters.

He said the bone-chilling effects of the lake didn't sur-face until he made it to shore.

"It was the last thing on my mind in the lake," Engst said. "But once I got to shore, I've never been so cold in all my life. I couldn't stop shaking."

Both men were rushed by ambulance to St. Joseph's Area Health Services in Park Rapids, where they were treated and released.

"It was a miracle. We're glad we're here. It could have been the other way."

those who helped in his rescue, including sheriff's deputies, police officers, first responders, ambulance crews and those along Long Lake.

"We're just so thankful that everything worked out," he said.

Engst repeatedly thanked

In the early winter of 1998, I was backing up my 4-wheel ATV with an attached trailer to load some furniture into the garage. My shift on the ATV malfunctioned and although it indicated reverse, when I pressed the accelerator, it sped forward & over a four-foot retaining wall.

I was thrown off the ATV and then the ATV fell on top of me, with the trailer on top of the ATV.

I was unconscious for some time, my head resting in a bank of snow.

When I regained consciousness, I was able to crawl out from under the wreck and inched my way back to the house. Mary finally heard me and called an ambulance.

I was airlifted to the hospital in Fargo, North Dakota where I had a severe skull fracture, loss of cerebral spinal fluid, and a few other complications.

But again, I made it.

So, in hindsight, were these really mishaps? Or more added blessings?

Around The World

I have been privileged to have visited all the continents on the world, except Antarctica.

I have been in the countries of Japan, Philippines, Vietnam, Australia, New Zealand, Mexico, Chile, Peru, Colombia, Panama, Costa Rica, Jamaica, Mexico, Canada, Egypt, Jordan, Israel, Turkey, Belgium, Netherlands, Luxemburg, Denmark, Norway, England, Scotland, Ireland, Germany, France, Italy, Spain, Poland, Czech Republic, Slovakia, Hungary, Austria, Bosnia, Croatia, Switzerland, and Serbia. As mentioned earlier, we have been to the islands of Guam, Wake, and Midway in the Pacific, and the islands of Majorca, Crete, Peloponnese, Cyprus, Rhodes, Patmos, Mykonos, and Crete in the Mediterranean.

Besides yearly trips to Canada for fishing and Mexico to work at the orphanages, we managed just about yearly travels to foreign lands with family and friends. By my recollection, Mary & I have now been to Italy at least 10 times, 2 while in the military and 8 after. We have been to Pisa, Venice, Rome, and the Italian Riviera multiple times. We also squeezed in two trips in 2006 and 2013 to the US southeast & southwestern states.

Most of our other foreign travels were with Lori & her husband, Michael Sullivan. Many of these trips, our daughter, Jackie also joined. These were memories of a lifetime.

We have been to all 50 states in our great country.

I will try and describe a few highlights that we experienced.

Germany

While still in the military, I was sent to Berlin, Germany on business. After WWII, Germany was divided into 4 zones of occupation, French, British, Russian, and American. The French, British, and American areas were considered part of the German Republic of West Germany while the Russian controlled zone was of East Germany. Berlin, the capital of Germany, along with a large surrounding area of Berlin were in the Russian controlled zone. Berlin was further divided into East and West occupations. The French, British, & American formed West Berlin & the Soviet sector, East Berlin.

Because of this unique arrangement, Americans like me could only travel to Berlin via established corridors from what was now West Germany into or out of Berlin. Escape to the West by German citizens from the East side got to be such a problem that the Soviets built a wall dividing the city, known as The Berlin Wall. It was heavily guarded by East German communist security forces and included areas of entrapment like barbed wire & even mine fields.

When I arrived in Berlin by duty train, I got to see the impact of Russian occupation on Berlin, firsthand. Our intelligence unit that I met in Berlin was garrisoned in the Olympic Village that was erected for the 1936 Olympic Games, formally called, "Die Olympischen Spiele 1936." The Summer Olympic events were held in Berlin and Winter events in Garmisch, amongst the beautiful Bavarian Alps of Germany. I got to take a dip in the Olympic pool that was still operational & I also got to see a lot of The Berlin Wall, including the historical "Checkpoint Charlie," the heavily guarded gateway of East & West Berlin, the Soviet and American sectors respectively.

After my military retirement, Mary and I were privileged to again visit Berlin, chaperoned by Lori & Mike. During this trip, we visited the Kunz family and celebrated the 100-year anniversary of their home!

Mary: We got a phone call from Helmut Kunz. He called and said he just got out of the hospital. He said, "Please, would you come over and visit?" Artie & I pondered over it and kind of decided that we couldn't at the time. When we told Lori & Mike about it, they said, "We're going!" They made all the arrangements, air travel, and car rental. After we arrived in Germany, they had a big 'Oktoberfest' party, and all the neighbors came by to also celebrate the 100-year anniversary of their house. We had a great time. Unfortunately, Helmut died shortly after we got back home. We were so happy that Lori

& Michael insisted that we go. Lori & Jackie are still in touch with Helmut's wife and family online & Artie & I are still in touch with our friends from Germany today.

By the time of this visit, all of Germany was reunified & The Berlin Wall had been dismantled, so we were able to spend most of our time in East Berlin, although it still felt like two cities in one. East Berlin was in a stage of extensive rebuilding & recovery, which indicated that it would soon be, once again, a vibrant "free" city, a testimony of many changes past & present, and the differences between communism & free market democracy.

Happy reunion with the Kunz family

Travels With Ron & Ellie Engst

Our first cruise line voyage was with our friends, Ron and Ellie Engst in 2005. We took the Inland Passage trip to Alaska and back. First, we traveled by Amtrak train from Detroit Lakes, Minnesota to Seattle, Washington, then by sea up the west coast of Canada & Alaska. The sea route kind of goes in and out of the islands that dot the coastline, thus allowing for very little ocean turbulence.

In Alaska, we had stops in Juneau, Skagway, Sitka, and Ketchikan. We saw calving at the Sawyer Glacier, learned history of Gold Rush days, and many fascinating facts of the unspoiled area of our country that is Alaska.

Another memorable trip with Ron and Ellie Engst was to Central and South America. We boarded a huge ocean liner in Miami with stops in Jamaica, Costa Rica, Panama, Colombia, Peru, and Chile.

Jamaica has some of the most beautiful shoreline I've ever seen. In stark contrast were the slums in the inland cities and apparent poverty of the people.

On our stop in Costa Rica, we took a train through the jungle. The train had to be at least 100 years old and driven by steam. The cars were comfortable but made mostly of wood and the windows were kept open during the excursion. We made a couple of stops at some small villages but otherwise it was solid jungle with all kinds of wildlife and insects. Our passage through the Panama Canal was a testimony of man's ingenuity and success against all odds.

The isthmus is miles of Cypress swamps with a series of connected lakes and rivers. The passage through the locks is huge and impressive. I cannot remember the difference between the sea levels of the Pacific Ocean and Caribbean Sea, but I think it might be around nine feet.

The shoreline of Chile is similar to California, fertile valleys lush with crops and vineyards abound. The backdrop is the majestic Andes.

We flew home from Santiago.

Our ship going through the locks on the Panama Canal

Puerto Vallarta is still our favorite vacation city in Mexico. We first went there with our daughter, Lori, but then the next time we took our favorite traveling buddies, Ron & Ellie Engst.

Ron and I tried parasailing! I was 75 years old. I loved it. I could do it every day as a matter of fact. Birds came and flew beside me. What an exhilarating experience.

Beautiful city, wonderful beaches.

Me, age 75, parasailing and landing on the beach in Puerto Vallarta.
Exhilarating!

Europe

Lori's husband, Michael Sullivan, arranged for us all a week's vacation in Eastern Europe. We stayed in a house that was 600 years older than America. Just makes you sit there and think, we really are a young nation. We're only 250 years old! Some of these nations have been in existence for a thousand years. Things like that grab me and put a better perspective on who we are, or how young we are as a nation anyway.

They had taken this old building, installed electricity with pipes all along the walls. The stairs were original, must have been oak, gauged down from people walking on them for years and years. You think about the people that lived in that building over the years, in this ancient country of at least a thousand years.

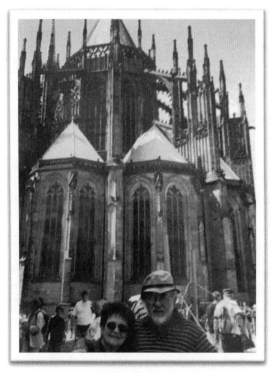

Mary & me,
St. Vitus Cathedral, Prague

One of the most challenging trips we experienced as a family was while we were stationed in Germany on our second tour, 1972-1974. Mary's parents and her brother visited us, and at the same time Lori brought her college boyfriend, Michael Sullivan, to visit. I had two cars at the time so, Michael drove our Volkswagen Beetle with all the kids, Lori, Jackie, and Tommy. I drove our Fiat with Mary, her parents, & brother, while also towing a trailer and camper.

With the two cars, we traveled through France and Italy.

A big problem occurred when we were in Italy. Mary's Dad drank unpurified water at a gas station and suffered a terrible attack of diarrhea. Fortunately, I was able to get him some medication at the American military base in Vicenza, Italy.

This was compounded by the fact that the campsite latrines had outdoor toilets, which consisted of a hole in the floor with two footpads beside the hole to stand on. Our family knew all about this from our days in Japan, but Mary's folks and her brother...let's just say they had a lot of trouble. This was already becoming quite the trip to remember!

Onward, we spent time in Venice, Verona, Florence, Rome, saw the Vatican, the Italian Riviera, Pisa, Livorno, and then went north through the train tunnel from Italy into Switzerland. As I recall, we drove onto a flatbed and stayed in our cars for the hourlong railway tunnel. The beauty and majesty of Switzerland is beyond words, like a fairyland.

Our trips to the Cinque Terre, Ostuni, and Pompeii, Italy are each stories in themselves. I believe our visit to Puglia and our stay in a trullo, the traditional Apulian stone house, was most memorable.

-Here goes.-

This trip started in Croatia. We first rented an apartment in Split for a couple days. A ferry took us to the Island of Hvar. We had beautiful rooms with a harbor view. From Hvar, a ferry took us to a bus station which then took us south to the city of Dubrovnik. An armed guard accompanied our ride & made us feel a bit insecure.

A small strip of land divides Croatia and is part of Bosnia-Herzegovina. On this border strip, an armed guard from Bosnia would switch places with the Croatian guard when we crossed into Bosnia, then another switch of the guard when we crossed back into Croatia. These countries were still recovering from their ethnic wars, distrust still abound.

One evening in Dubrovnik, we were walking in the street, and a chef shouted over to us,

"You people looking for a place to eat?" He said it in pretty good English, too.

"Yes!…You got good lamb?" I was kind of joking, but then,

"Oh, we eat a lot of lamb over here! Why don't you come in, your whole family, I've got a room down there for all of us. We'll sit down and have a real meal together."

He made us a full, complete meal, what a Croatian family would eat on a special occasion. Not off the menu either, all the goodies included. From the first serving to the last, including a special wine that they only use for special occasions, we ate, talked, and laughed. His son enjoyed practicing his English with us, too. What a great, memorable day.

Private dinner with the chef!!

After our visit to Dubrovnik, we took a ferry across the Adriatic Sea to Bari, Italy. This area is the heel of Italy geographically, and the village of Ostuni is located close to the Adriatic on the east side of Italy. It lies between the seaports of Bari and Brindisi. Our luck would have it that the Italian Air Force had now taken over a former US Recreational area. Luckily, I had taken my military ID card on the trip which gave us access to a beautiful beach area with umbrellas, a beach house, restaurant, you name it. We went there nearly every day. We rented a traditional stone cottage, a trullo, with a cone-shaped roof that was situated in an olive grove, filled with fig trees. These stone dwellings, trulli (plural), are scattered around the regions of Puglia. They were often the homes of farm laborers, with the highest concentration of these homes in the town of Alberobello.

Our trullo home on a fig & olive grove outside Ostuni

The family in the town of Alberobello
Mary, Jackie, Mike, Lori, Me

We've been to England and Ireland on a number of occasions. On one trip, we spent a week in the London area. I had a chance to visit all the usual tourist spots. My favorite was The War Museum, which covers the history of the British Empire and their vast control of much of the world. I was also impressed by Churchill's War Room, the underground headquarters during the early years of WWII. It's located in the subway system deep under London itself. All the equipment, war rooms, and sleeping areas are preserved just as they were. Fascinating, don't miss it if you go there.

Our visits to Ireland were centered a large degree around the ancestral area where Mike's family came from, primarily County Cork. The old farmstead of Patrick Sullivan is still in the family. In fact, a cousin, James O'Regan owns it and provided a guest house for us during one of our stays. Unfortunately, we never got to Belfast in Northern Ireland. That's where my Grandma Burns' (Walston) family emigrated from during the height of the potato famine.

The breakfasts in Ireland are the best I've ever experienced – bar none.

In 2015, Lori and Mike invited us to go with them on a Danube River cruise. They planned it to occur in the late fall. We left from Budapest, through Austria, and ended in Nuremburg, Germany during the traditional Kris Kringle Markets, when large towns have a market in their town squares, selling Christmas ornaments & crafts. The real highlight is the food stalls. Roasted bratwurst, mulled wine (glühwein), rum punch, toasted nuts, and wonderful holiday gingerbread cookies (lebkuchen). The aroma of these stalls on a cold December night is simply an experience that is hard to explain.

On our stay in Vienna, Austria, we attended a performance at an opera house. Awesome! A real "Biggy Christmas present!"

Kris Kringle Markets
Nuremburg, Germany

Mediterranean

A couple years later, we flew to Athens via London. We visited the Parthenon, Mars Hill, Olympic Stadium sites, and Delphi, an ancient site, the center of the gods from Greek Mythology. We also visited the Mediterranean islands of Rhodes, Mykonos, & Patmos.

We also visited Israel and Egypt on a cruise. Our time in Israel, Egypt, and the Island of Patmos were the highlight, with the days spent in Israel & Egypt being the capstone of this voyage. The Cave of the Apocalypse is on Patmos where *The Book of Revelation* was written. An interesting statement by our tour guide,

"All the inhabitants of the island were converted to Christianity, including the Roman Garrison before John was freed and went back to Ephesus where he spent his final Earthly days."

We visited Bethlehem, worshipped at the 'Wailing Wall,' and visited Jerusalem on foot. On the Mount of Olives, overlooking Jerusalem, there was a little plot of olive trees that they believe were there at the time of Christ, as seen below.

Egypt was a kind of optical confirmation of the once mighty empire. The pyramids and related monuments to their pharaohs and their families of that period of history are impressive to say the least.

The highlight of our Egyptian visit, however, was a trip down The Nile on an observation barge. We were escorted by two armed security boats. We went by the US Embassy, which, supposedly, is where Moses was placed in a floating basket and Pharaoh's daughter rescued him and was then raised in the royal household.

Our stop in Istanbul (Constantinople) was an experience in itself. We visited the Hagia Sophia which is one of the oldest and maybe largest Christian Cathedrals in the World 360AD. This city was the Eastern capitol of the Roman Empire. When the Islamic Ottoman Empire conquered this part of the World in 1453, they fortunately didn't destroy this magnificent building, but just hung Islamic murals over Christian ones. It is now a museum.

This area is where Asia & Europe meet. In fact, there is a place on the isthmus where you can stand with one foot in Asia and one in Europe. So, one can say they've been in Asia and Europe.

Australia & New Zealand

Our most unusual trip was to Australia and New Zealand in 2009. Mike, again, planned most of the trip. We stayed in Sydney for a few days and visited the usual tourist sites, like the Opera House. A real treat was that I was able to get in contact with the Australian Intelligence Representative that I worked with on my first tour in Vietnam. He represented the Australian forces in Vietnam that were in the same AO that we served in Central & Southern Vietnam. Super guy!

We drove to Cairns for a few more days and were supposed to go out on the Great Barrier Reef that is located off the northeastern quadrant of Australia. The largest monsoon to hit the continent in years hit on our day, so no chance, as nobody went out to sea that day.

However, we were able to take a long cable car trip over the jungle rainforest area to a mountain city where the American Army had one of the largest hospitals in the Pacific during WWII. If I ever had to live in Australia, that's where I'd go.

Our four-day stay in New Zealand was simply put, four days of heaven on Earth. Maybe I wouldn't do the mineral mud baths again, but the weather, people, and scenery were just spectacular. The air flight over & back is long, in case you ever plan this one.

Sitting with Mike near the Sydney Opera House
Sydney, Australia

Skyrail Rainforest Cableway
Cairns, Australia

While writing these memories I was reminded of a part of our family life that, I'm sorry to say, is vanishing. Mary and I were fortunate to grow up in a time where at least two meals a day, at the table, was a daily ritual. It was a time to share conversation and bond. I found that sense of presence in most of the countries we went to. When we traveled in Spain and Italy and in all the Mediterranean countries, that's how they dined, gathered around with others, enjoying that moment in time. That was something I loved about 'culture.' I wanted to get in with the people, not just sit on the military post. And it just happened that way. That wherever we went, we stayed with families. Even when we camped, we were invited to other campsites to enjoy a meal with conversation. It was really fun to exchange fragmented pieces of language to get all our points across.

Mary: If you get the chance to travel, do it.

FAMILY THOUGHTS

Our Military Lives

Mary & Me

Mary: I was always very proud of Artie. I was very busy during those traveling years. And the times that I was separated from Artie during combat tours were traumatic. I'm very happy that we were able to travel and go to different places. I think it's made a big difference for my children. They are able to branch out, they're more adventurous, and they're stronger. I appreciate it all.

And Mary is right. They are also more confident. They are self-assured that with anything, they can do it.

Mary: Jackie adjusted the best. Lori rebelled every-once-in-a-while and Tommy just went along with it.

Mary & Me
& our children, Lori, Jackie, Tommy

Through Lori's Lens

"The world is a book, and those who don't travel only read one page." – St. Augustine

I have to say being a military brat, the benefits outweighed the negatives. It really was a rich life, full of unparalleled adventures. It offered a way of seeing the world that not many of my peers got the chance to experience. We lived a gypsy life, military nomads. We were a unique group of kiddos.

It was easier living near military bases. We went to school with other military brats and we had teachers that were ready for us. They understood that they had a great population of transient students that changed school systems from all over the US and World. They expected our frustration and anxieties. They *got* us. Every change that we had to make came with the trepidation of, *ugh, I have to be the new kid again.*

Being a military kid & during war years meant that you had to leave your dad and be somewhere else without him. You were not tied to a military base for the most part and you often had to be home with other family members. Those situations were far more traumatic for me than anything else. It meant having to go to a place where everything was so established and people knew each other their whole lives and you bust into this steady society, which would easily and quickly turn into the hardest place to be.

My sister, Jackie, life was a big adventure for her. She loved it! She embraced it 100%, always ready to go, and eager! I was the one that was more sensitive to the changes, the designated worrier of the bunch. But was it good for me? Absolutely.

With each move and each experience, we saw the world differently, and the world would get smaller and smaller.

I really hit the jackpot with my parents. Both had the spirit of adventure and were shoe-ins for military life. They embraced change.

My mom was amazing. Often, we arrived at a new base with nothing, but she had the prowess to pack the bare essentials. She was the organizer & financier of all the travel when we were young. She saved grocery money during the school years to use for travel in the summer. And camping was the most economical.

Every family trip we took, Mom was copilot with the maps. Every time we stopped for gas, mileage was calculated & recorded, dad's penchant for analytics.

My favorites were the beach vacations. We stayed 'til the funds ran out. We had to get used to iodine in Tang, a way to purify our drinking water, but it was a small price to pay. After all, we camped throughout Europe!

It's funny when I recollect our camping experiences. The Europeans had camp gear that was so bright & compact. Everything was lightweight and folded down to neatly fit in a Volkswagen Beetle. We had a big station wagon stuffed with army gear, heavy canvas tents, big clunky coolers, and Coleman stoves. It was a sight to see! I think we were a hit because we stood out!

We met an Australian couple camping who later showed up at our door in Germany and stayed for a week. My parents' home always had an open-door policy. Guests came from all over the World. It is still like that today. Mom loves to entertain. I am in awe of my mom and how she kept everything running so smooth while we lived a life of transition.

I loved our time in Germany. I loved Kathe Kunz, the master of kirsch kuchen, and her sons that welcomed us!

Dad was the fun guy! On the beach he would wake us up for a swim before breakfast. In Barcelona, we watched in fascination as he ate fresh, spiny, black sea urchins that the spear fishermen caught. He really has stayed young in spirit his whole life.

Cryptograms & crossword puzzles, he has always loved puzzles of every single kind – Wordle, now. But all his life he's been with a puzzle. Mom goes shopping and he finds a nice comfy chair. Analytical to his core.

The military life prepared me for my life with my husband, Michael Sullivan! He got the 'travel bug' the first time we visited Germany when we were college students, before we were married. After that, Mike's trips that he planned for us were epic adventures! Always 'the road less traveled.' We slept above a bar on straw mattresses, with roosters to wake us in the morning. We dined with olive grove owners, who jokingly said they were in the "oil business." We laid on a beach to watch a meteor shower and took mud baths in New Zealand. We visited a wine co-op in Italy that dispensed the local red by the liter via gas pumps and played bocce ball with the locals in a plaza in

Switzerland. These trips were planned every year when our children were older and continued 'til Mike passed away. My parents and sister joined when they could. I thank Mike for these wonderful times.

I had the chance to travel to Vietnam. It was wonderful to see the place where my father invested so much time. It would probably be hard for him to see the old base camps as some have been converted to museums that now paint an evil picture of the United States.

I was in my 40s when I hosted Russian exchange students & my parents paid a visit and got to meet them. Dad saw some of their Russian writing on the table and started translating! We were all stunned.

I never knew what my dad did. As far as I knew, he was in the Army. He was always a quiet man. We never asked him, "What did you do at the office today?"

When he went to Vietnam, he assured us he just had a desk job, no worries.

This stunning revelation, when I learned that my dad knew Russian, was the inspiration for this book. I wanted to know more about his life.

Dad lived through exciting changes in this country, but also was witness to the shady underworld of the "Cold War." I never knew the extent of his involvement. I knew he had medals and decorations. But all of this? This has been an even bigger adventure than I first thought. It's been great. It's been profound. It's even bridged the gap of how I felt about Vietnam & how each of us was impacted by it.

I want the future generations of my family to hear their grandfather's story. I want them to understand that this is a very small planet inhabited by wonderful people and cultures.

Wouldn't it be wonderful if every child had a chance to live for a year in another country? If at all, it would go a long way to improving foreign relations.

So, here's to you, Dad! Thank you for your guidance, love, protection, and mostly, the adventure!

An epic trip to Long Beach, South Carolina

Dinner with the "oil men"

Wine co-op
Manduria, Italy

Through Jackie's Lens

I was in third grade when Dad was in Vietnam for his first tour. I was caught up in the excitement of living with Grandma & Grandpa and wasn't too concerned with the war. It became real to me when the news showed a Vietnamese man with his hands tied behind his back, shot live on TV. I just happened to be alone in Grandma's living room when it was televised. In later years, I heard that it was the network's mistake to show that footage.

It was great living in Park Rapids. Grandma and Grandpa spoiled us! It was also a treat to go visit Grandma and Grandpa Burns' farm.

Despite the circumstances, I never felt depressed during that time. We got a lot of support from everyone.

During that year I was going door to door selling all-occasion greeting cards for extra Christmas money. Not that I was scheming or anything, but I told them that my father was in Vietnam which helped elevate my sales!

Sometimes Mom wouldn't hear from Dad for long periods of time and then she would get a bunch of letters all at once and disappear to go read them in private.

I remember becoming fearful when we would spot a military vehicle in town. If your loved one was killed in the war, they would come to your door to inform you.

Mary: One time that really frightened me, I was taking care of my cousin's dogs at her house. I looked across the street and this military car was driving up. I just knew, I had this feeling that they were coming to talk to us. But they went to the neighbor across the street. And I just stared out the window and watched. And I watched when she shook her head "no" and pointed over in my direction.

I remember you crying, Mom, telling Tommy & me, "Don't answer the door!"

Mary: I called Artie's Father to just come and sit with me, I was so frightened. Later, my cousin went over to that house next door, who was just saying that her son was AWOL & that she did not know where he was. It looked like she was pointing to us, but she was not. They never came to our door.

Her cousin, Sharon, would come over and tell Mom, "You're not gonna sit around and mope, (which she didn't). You're gonna LIVE! Come on, we're going." She'd take her somewhere and it just helped to make us all happy during the harder times. Sharon was "the best!"

Growing up in a military family grooms you for real, adult life. We learned how to overcome challenges. We learned sacrifice because boy, what we sacrificed to live the life we did. We went through some hard times. Like, losing my dad for a total of three years. It was our duty to step up and help Mom, to roll with the punches of our lives and contribute where we could, like doing our chores without whining. You couldn't be a brat, which I was sometimes, and that's the truth. Sorry, Mom.

My brother, Tommy, was the introvert, with an inventive mind and a beautiful soul. And my big sister, Lori, what a beauty queen! Mom & Dad didn't ask us to look out for each other, but we did anyway.

At Christmas and Easter, we'd put together boxes of fun 'gifty' things for Dad and his friends. I was in charge of buying the lime suckers for Dad. It was his favorite flavor.

I remember the occasional phone calls to Dad at Christmas or Easter. It was only 2-3 times a year that we got to talk to him.

We lived in Park Rapids during Dad's second tour in Vietnam. It was the second half of 7th grade and then through 8th. Mom and Dad lived in a huge Victorian home that they made into apartments.

We moved to Germany when he came home. Lori stayed in the states and went to college. So, it was the first time without my older sister.

Going to Germany the second time, OK, I wasn't *as* excited, as you can imagine, at my age. I was going into 9th grade. Dad had just come back from Vietnam, and I was excited to be with him, at home, in Park Rapids, *and* have my friends, too. These friends I had formulated over 18 months, *and* I had a boyfriend!

Times were never dull. We *never* sat around and did nothing. They took us everywhere. And I appreciate that now. Mom saved up every penny for us to go camping & travel all throughout Europe. We explored castles and camped in Spain. We went to Italy, my favorite, my favorite, *my favorite.*

I remember being excited that Lori was coming to visit and she was bringing her new college boyfriend, Michael Sullivan. Something new, woohoo! I could not wait for whatever experiences and adventures he would bring, and I didn't even know him yet. Grandma & Grandpa were also coming over with my Uncle Wally.

And I hit it off with Michael Sullivan immediately. He fit with our family seamlessly, a gypsy spirit about him, a desire to wander, to explore, and have adventure.

So here we were, our big family in Germany. We took off in a Fiat and a Volkswagen bug. We had a trailer tent that opened up into a double sleeping bed arrangement. There were 9 of us! Driving through Germany, down through France, and the tunnels that took us through the Swiss Alps.

And then our Fiat broke down. In the mountains…in Italy…in a tunnel. Luckily, we were able to pull off the roadway a little because, you know, everyone drives at least 100 miles per hour. Here we were, this American family. We were all standing around with Grandma, Grandpa, & Uncle Wally, outside of our cars. And guess who happens to be coming through the tunnel? The company maintenance man for Fiat that monitors the autostrada (highway)in Italy.

The Fiat guy got out, helped us, knowing exactly what to do. Fiats tended to lock-up, I guess. He got his funnel and poured cold water down into the pump as we all stood around and watched. Within 20 minutes, away we went, sailing along like nothing ever happened.

We camped all throughout the trip. My family, jam packed into those two vehicles. It was just so awesome. I'll never ever forget camping in Switzerland. We got there late at night, so, so tired. I don't even know how we pitched our tent, it was so dark. When I woke up, I mean it, I thought I was in heaven. Unbeknownst to us, we were on top of a mountain, looking down below onto a beautiful lake, white swans swimming. Grandma and Grandpa were in disbelief and awe. Grandma said, "I think this is what heaven is gonna look like."

One of the two stove burners went out while we were there. I remember Mom cooking for all 9 of us on that one burner. The storekeeper from our campgrounds brought around hot and crispy brötchen (bread) for us. Anytime we took a road trip, Mom would pack a cooler with bologna sandwiches, and it would drive Lori insane! I mean, she would go ballistic.

Mary: Whenever we'd stop, we'd say OK Lori, get the cooler! Oh, she would be so upset…

Arthur: I would always tell her, you can have the big sandwich!

Lori was a little gypsy too, just like Michael. Pitch a tent? She wanted a hotel room! So, she would sleep in the car, protesting the tent. Or we'd be driving around, I'd be saying how much fun I was having, and she'd hit me in the arm. She's funny like that.

Michael had sprained his ankle, so he limped around the whole trip. Grandpa drank the water in Italy and had terrible dysentery. We had to go to an Air Force base to get him some medicine. Mike & I drank too much Sambuca, because in Europe, you can drink at any age with your parents' permission…or without.

Dad had us on 'KP,' kitchen patrol, throughout the trip, organizing duties like we were his own little platoon. Lori washed dishes, I dried, and then we'd rotate. Dad ran a tight ship, but it was good for us.

In Heidelberg, I had this friend who loved to come over to our house. She loved coming over for dinner and loved spending the night. *Why?* I always thought, *when we have a curfew, a bedtime, rules, all that stuff?* Well, she loved that because she didn't have that at home. She thought it was great that my parents cared that much for us to have a curfew.

In the military, you don't want to get into too much trouble because that was a direct reflection on your father. He could get talked down to from his superiors and we knew, innately, that we didn't want that as a family, as his kids.

So here I was, seeing the world through the eyes of a teenager. It was an absolute adventure. We were always going on adventures, always living on the edge somehow. We had packed up our lives in the United States and went overseas and now into multiple foreign countries.

I was worldly. I was exposed. To the good, the bad, the ugly, everything. I learned how to adjust and how to not be afraid. To this day, I'm not afraid of traveling or moving. When I first met my husband, Jeff, he was shocked at how I could just go online and buy us a trip and say, "OK, we're going!" "Aren't you apprehensive?" He would say. And me, "Nope. It will all work out." Mike Sullivan was like that too. We were great travel partners in that way. He was my brother. We are all teachers to someone in life. He was ours and we were his, just in different ways through different times in life.

I joined a co-ed explorer troop, and we traveled everywhere throughout Germany. We'd go on trains, go camping in Italy, & when we went up to Amsterdam, we stayed at a youth hostel.

I was just a teenager and absolutely loving my life. I became resilient, flexible, and adaptable. Every time I moved and attended a new school, I was able to reinvent myself. If I didn't like the direction my life was going, I could have a clean slate when I got to a new place. I'm so glad social media didn't exist in my day!

That's what it was like, for me, being in a military family.

My life was fun! Sure, it had its hard moments, but it was full of experiences and adventure. I was lucky to have my mom & dad look after me in the ways they did.

To my children, grandchildren, and great grandchildren: do not be afraid to live your life. Make the best choices that you can but forgive yourself if you don't. Reinvent yourself if you need to. We are supposed to go out into the world and explore it. It might be uncomfortable, but you're supposed to be uncomfortable at times. If the military taught me anything, with the uncomfortable comes the BEST of everything else. *Make the best of wherever you are, whenever you're there.*

I am grateful for my military life. It shaped me! What a wonderful experience. And I thank my mom & dad for that. And of course, I thank the guardian angels that have kept us safe all these years.

Greece!

227

Through Tommy's Lens

I was the youngest. I do not recollect much of the earlier years. I remember the second tour of Germany. A grand trip to Italy & Switzerland with my mom & dad, sisters, Grandma & Grandpa Johnson, Uncle Wally, and Brother-In-Law, Mike, happened the summer of 1974.

I slept under a camper in Livorno, Italy. We had heavy rain one night and I woke up to a very wet sleeping bag.

Our family made it to Pisa and I went with my dad and Lori up to the top of the leaning tower. Halfway around, you seemed to be falling up the stairs, and the other side, to be crawling up the stairs. At the top of the tower, you could stand directly under the bell and touch it. I remember there weren't railings on the edge. They certainly wouldn't allow that today.

On an earlier trip, our family arrived late to a campground in the Mediterranean and had to wait 'til morning to check-in. We temporarily setup our pop-up trailer on wheels for the night in the parking lot. There was only room for Mom & the girls, so Dad & I waited for them to rest a little and watched the bats flying around. Eventually, Lori came out (claustrophobic) and I got a bit of sleep in the camper. I remember driving our car onto train cars that went through a mountain pass. It was so narrow that if you put your hand out the window, you could lose it.

I remember being too short to see the Statue of Liberty as the ship came into New York harbor on the way back from our first tour of Germany. Someone finally lifted me so I could see.

I remember living in Maryland before Dad's second tour in Vietnam. We lived in military housing on Lawson Loop. One of my most vivid memories of that place was watching the first landing on the moon. The power went out in our area. Mom and Dad invited many neighbors over to watch on a newly purchased, tiny battery operated black and white television.

In Germany, we had to use large heavy boxes that were converters for any electrical appliance made in the United States. On the plane ride back to the US with Jackie, I enjoyed the first airplane video screens with a movie.

Sometime after high school, I followed in my dad's path by enlisting in the military. I went to basic training in Fort Sill, Oklahoma. Around Christmas time, I surprised the family and hopped on a plane to Minneapolis. Mike & Lori were so surprised and quickly came to get me. They called Mom & Dad and made up some reason for them to come down to the cities so I could surprise them.

Recently, I needed to apply for Social Security benefits. I ran into trouble because of my foreign birth. Apparently, some paperwork was incomplete from 1959 when I was born in Japan. Luckily, Dad found my original birth certificate that proved I was born in an Army hospital and that I was not a citizen of Japan. A Social Security application was missing from their files.

My greatest memories are all centered around the travel we were able to do while growing up, thanks to Dad.

My Family
Thanksgiving 1992

POSTSCRIPT

I lived in some remarkable times and man, things happened to me that were unordinary in some cases, maybe weird in others, or simply just not the norm. And yet, it was a good life. Of which, Japan was probably the most shocking yet exciting. But I looked forward to each and every experience and I can pleasantly look back on 'em, too.

I was fortunate enough to be able to choose a vocation that I enjoyed, and I thank God for that. I believe there was a hand in that and a mighty hand through some very troubled times, moments, injuries, and blessings.

Today, I could very well be a vegetable. I could be paralyzed, a paraplegic, but I'm not. I feel good, actually. Maybe I don't look so good and maybe my mind isn't as sharp as it was at one time, but by in large, no pain, and I can remember people and things. So, I'm happy about that. I'm thankful is the best way I can put it.

To my offspring, I pray that these few memories of my life and family will be an inspiration to believe in our Lord and Savior and live your lives to the fullest. Don't lose the faith and don't lose your hope, in the good things of life now and the life to come.

And to be honest, Mary, Lori, Jackie, Tommy, I never really knew how you felt about certain things, our lives, & our travels, until the making of this book. It's been great to learn & to hear from you now, firsthand, of your own unique experiences & challenges.

I thank God that I was able to leave the farm life and travel and see the world. To the farmers, past, present, and future, I honestly admire and appreciate your chosen occupation and achievements.

A lot has happened in this world since October 1930. I would be the first to admit that there are a few events in my life that I wish I had done differently, but who can't say that in retrospect? I am satisfied with my time thus far on this ever-changing planet.

All considered, I am at peace with my God, country, and my testimony of life. I thank God I was born in the United States of America. My parents loved me & supported me to the best of their abilities, faith, and knowledge. I was blessed with a wife that loved me, supported me, and avidly enjoyed our travels and loved our children with a passion. What more can a man ask for?

That's about it, family, friends, and readers. I pray you enjoyed my journey.

The Shalom of God be with you all!

Arthur Lee Burns

CO-AUTHOR'S NOTE

To Arthur, to Mary, thank you. For sharing this life that you've lived. For sharing your home with me in order to collect these stories. To allow me just a glimpse into your life, a glimpse far apparent as to where this story would lead. One of love & adventure, resilience & sacrifice, family & gratitude.

To Mary, for afternoon ice cream cones, home cooked meals, & freshly brewed coffee with your secret ingredient. For late night snacks that fueled our chats far past all of our bedtimes. And Arthur, grabbing a bottle of red wine from the patio, naturally chilled from a small bank of Minnesota winter's snow. With you & Mary, it was like sharing stories over a campfire. I couldn't describe a more warm and welcoming moment in time than it was to put together this collection of stories with you.

Thank you for showing me around Park Rapids, into town, to the nearby farmland, & into Osage. For allowing me to enjoy your family, friends, & frequented places. For completely bringing your stories to life right before my eyes.

Thank you, Lori, thank you Jackie, for being my eyes & editors. For sharing the lives *you've* lived through your own lens', offering our many generations of readers with a three-dimensional view of the world you were blessed to experience & the human emotion felt behind the scenes.

To our readers, I hope you feel enveloped by inspiration. I hope your eyes have been opened to the possibilities that come from striding forward into life, right into the unknown, knowing that it could give you something you never imagined possible. I hope that in these few pages of time travel, you feel compelled to expand your own level of worldliness, by learning or traveling or conversing with worlds different from yours.

I hope that you have become more enriched by the history behind you. I hope you feel more understanding of what it takes for the freedoms you get to enjoy & every single life that has played and will play their part in that.

I hope that you choose to pull as much adventure from adversity, leading your own life with immeasurable courage & willingness toward experience for your own collection of stories.

Monica Marie

Me, Monica, enjoying a noon ice cream cone with Arthur & Mary

The view from home,
looking out onto Fishhook River
& the American Flag

Made in the USA
Columbia, SC
13 December 2024

a6a89a29-a895-4b06-98a1-bc9ebeb6420cR02